Roxbury Community College
Science Department

SCI 124 Principles of Chemistry II
Laboratory Manual

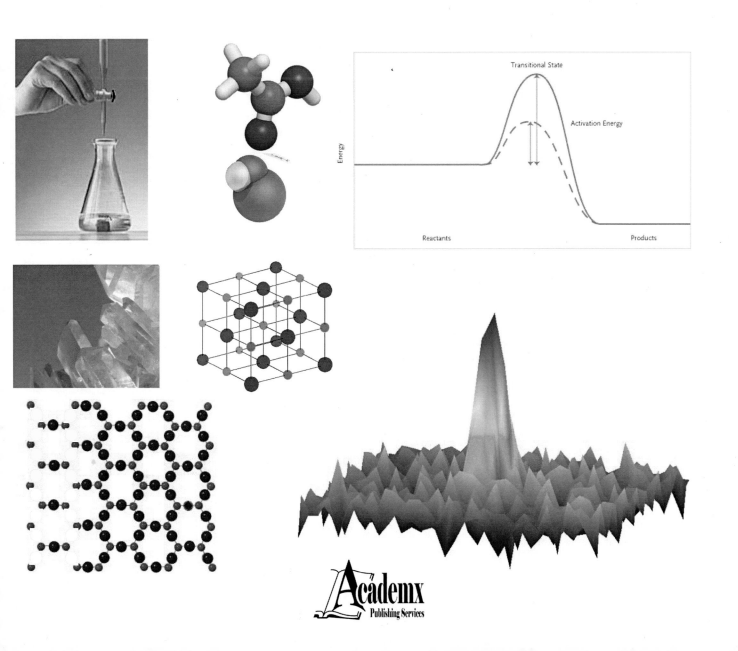

Academx
Publishing Services

Acknowledgements:

Bill Griffin from Bunker Hill Community College kindly provided the electronic copy the original template from which many of laboratories were developed and has given permission for us to reproduce many tables and sections of this manual. Ching Yim from Roxbury Community College wrote Laboratory 10 Periodicity. Andrella King spent a great deal of time optimizing the Iodine Clock Reaction (Kinetics Lab) at RCC. Rajeswari Sundaramoorthi (RCC) provided a very clever solution to the high error obtained in the original electrochemistry laboratory. Special thanks to Kenneth Hede from University of Massachusetts Boston for helping us with text editing and preparation of prep sheets for our technicians to set up the labs. Others have given many suggested changes in the details of many these laboratories. Suggested changes are always welcome for an improved edition at the start of the next semester.

Principles of Chemistry II Laboratory Manual, Second Edition – SCI 124, 2014 - 2015
Copyright © 2014 by the Science Department, Roxbury Community College

All rights reserved. No part of this publication may be reproduced or transmitted in any form or by any means, electronic or mechanical, including photocopying, recording, or any information storage and retrieval system, without the written permission of the publisher.

Requests for permission to make copies of any part of the work should be mailed to:

Permissions Department
Academx Publishing Services, Inc.
P.O. Box 208
Sagamore Beach, MA 02562
http://www.academx.com

Printed in the United States of America

ISBN-13: 978-1-60036-795-3
ISBN-10: 1-60036-795-X

TABLE OF CONTENTS

Roxbury Community College
Science Department

Chemistry Laboratory
Safety Guidelines

The Chemical Laboratory is a place which has more potential danger than most other locations on campus. It is important that everyone be aware of their environment and to work in a safe and responsible manner. Working safely in a laboratory environment depends on three major factors:

1) Using your common sense.

If you think something appears to be dangerous or unusual ask the instructor to check your workplace. In all cases follow your instincts and keep yourself from harm.

2) Following the directions of the instructor and staff.

If the instructor or laboratory staff of RCC gives you a direction or order you should follow their instructions immediately in what you would consider a safe and prudent manner. If you are told to evacuate the laboratory do so quickly and in an orderly manner by leaving through the primary exit (the door you enter the lab from).

3) Being careful and alert to what is around you.

You are surrounded by other students normally performing the same experiment as you are but in a different location. Be aware of what other people are doing as well as observing your own experiment. Does something seem very different with your experiment compared to others? Do you see something that might not be safe? If there is a problem only 3 feet from your location it is likely to affect you -- so, talk with your classmates or the instructor if something seems out of place (following #1 above).

Roxbury Community College Chemistry Laboratory
Safety Guidelines

In addition there are other rules and regulations that will ensure a safe and productive semester of laboratory work. Those rules are as follows:

- In the case of an extreme emergency have the instructor or laboratory staff goes to security or pick up the phone in the laboratory. Students should also pick up the lab phone for security if the instructor or staff is unable to call. Do not dial 911.

- There is no eating or drinking in the laboratory.

- Goggles are provided (which ensures they are chemically resistant) must be worn at all times after any pre-lab lecture provided by the instructor.

- Only water goes down the drain! There will be a waste container for all materials used in your experiment. If the waste container is not obvious ask the instructor.

- Dispose of all glass (broken, disposable or damaged) in the broken glass disposal box in the front the laboratory (not in the trash).

- If unsure about a procedure ask the instructor before continuing an experiment. Even the order of mixing some materials may be important. Follow the instructions from your instructor or the laboratory staff in the case of an emergency.

- Be aware of the primary and secondary exits that should be used in an emergency (fire or chemical spill). Always know where these exits are relative to your workspace.

- Know where the safety shower and eye wash stations are relative to your workspace. The shower is next to the front door in the front of the laboratory; the eye wash station is to the left of the sinks in the center of the lab. In the event of a major chemical spill which gets on your clothing you will have to remove that clothing and use the safety shower. Small spills which do not contact clothing can be washed off at the sink areas in the laboratory.

- Any materials which are expected to generate a harmful or noxious gas should be used in the fume hoods in the back of the laboratory. If your experiment unexpectedly starts generating a gas which you feel is harmful transport the experiment to the hood if you can do so safely and notify the instructor.

- If you have a fire at your workspace remove all flammable materials, shut off all natural gas supplies and smother the flames if possible using large beaker or other object. If the fire is substantial move away and call out to the instructor. There is a fire extinguisher near the prep room door but it should be used only by qualified personnel.

Roxbury Community College
Science Department

Chemistry Laboratory
Safety Agreement

As a student enrolled in a chemistry laboratory course at *Roxbury Community College:*
- I agree to observe all safety regulations as listed in the lab syllabus, lab manual and supplemental handouts.
- I have witnessed and understood the safety instructions and procedures presented today by the instructor.
- I agree to follow all verbal instructions of the instructor and staff and **agree to wear goggles given to me by the instructor at all times during the laboratory session unless specifically told by the instructor that they are not necessary**.
- I understand that if I violate any safety regulation, I may be asked to leave the lab and will accept a zero score for that lab exercise.
- I acknowledge that continued disregard for safety on my part will result in my being dropped from the course.

Principles II Chemistry and Laboratory   Fall   Spring   Summer   20 _____

Course Number   Section Number

_____

Instructor

_____

Name of Student (printed)

_____

Today's Date   _____

Signature   _____
Please pass in to the Laboratory Manager or Instructor

SCI 124 PRINCIPLES OF CHEMISTRY II
PASTE YOUR LABORATORY SCHEDULE HERE

SCI 124 Principles of Chemistry II
Laboratory 1 Experimental Determination MgO Heat of Formation (ΔH_{MgO})
(based in part on "Chemistry with the CBL" by D. D. Holmquist, J. Randall and D. L. Volz, Vernier Software 1995)

Purpose:

This laboratory exercise is centered on making measurements relating to heat changes that arise from chemical reactions. The enthalpy changes of exothermic reactions will be calculated from measuring the heat given off in these reactions. Hess's Law will be used to calculate the heat of formation of magnesium oxide (MgO) using experimentally determined enthalpy values as well as one reference thermochemical equation. The student will also determine the accuracy of their experimental data by comparing the calculated MgO heat of formation with the literature value of –601.8 kJ/mole.

Introduction:

There are many chemical reactions that generate or consume heat as well as form products. We are dependent upon the heat liberated in the combustion of coal, gas and oil to keep us warm in winter and to supply us with most of our electricity demands (stereos, computers, air conditioners, etc.). There are other cases where the heat generated or consumed due to a chemical change might be observed in our daily lives such as the application of a "cold-pack" to a sports injury or the use of "hand warmers" in cold weather activities. The cold-pack works because a concentrated solution is contained in a separate pouch within the pack and when burst is diluted in water. The heat of dilution (not a chemical change really but there is an enthalpy term for dilutions) is endothermic or positive and as a result absorbs heat from the environment. Something that absorbs heat from the environment will decrease in temperature since the quantities of materials have not changed but the heat content (enthalpy) is now negative and absorbs heat externally. The opposite case is true of hand warmers where exposure to air results in a chemical reaction that is exothermic (negative) and generates heat. This is an example where the reaction has a negative enthalpy change (heat is released from the system) and as a result the temperature increases or the hand warmer heats to temperatures above room (or outdoor) temperature.

In order to experimentally determine an enthalpy change in the laboratory, you have to be very careful in how you define "the system" and "the surroundings". The system is that part of the universe or the lab that you are going to measure while the surroundings is everything else (in the universe!). Between the system and the surroundings is a boundary since you will only measure <u>what exists in the system</u> and will not account for what happens to the surroundings. As a result it is imperative that you understand what the system is that you are measuring and where the boundary between the system and the surroundings exists.

In our case this distinction is fairly straightforward since we will be using nested coffee cups as a solution calorimeter and the **system is the calorimeter and everything that exists inside the calorimeter** and the **surroundings is everything outside the calorimeter**. We will have a pair of coffee cups (Styrofoam which is a good insulator of heat) nested inside one another, a coffee cup top and a glass tube containing a thermocouple for measuring temperature. Furthermore this thermocouple will be connected to a graphic calculator that is capable of recording temperature vs. time data for us to use in our calculations. In order to calculate our results correctly we will need to account for all heat changes within the system (calorimeter). If we have a reaction that generates heat in a solution then what gains heat and what loses heat? If the solids used are consumed (are not products) and that the solutions used are mostly water--is this true? What is the percentage of water in these solutions? We can make some simplifying assumptions.

The simplifying assumptions are as follows:

1. The room temperature does not change greatly during the experiment.
2. The only source of heat generated is from the chemical reaction being studied.
3. The heat generated by the chemical reaction is absorbed by the other components of the system which we can assume to be the mass of water present and the calorimeter (Styrofoam cups). We make the further assumption that even though in this case the cup contains hydrochloric acid, we will assume that this solution (which is mostly water) will have the same heat properties as pure water that is well documented.

If we use the symbol q for heat and H for enthalpy we can write some equations for an exothermic chemical reaction. If we assume that the calorimeter is well insulated then the heat from the reaction will be absorbed by the water and the calorimeter (and no heat is released to the surroundings). Where q_{soln} is the mixture of water and solutes (in this case Mg or MgO solids).

$$q_{total} = 0 = q_{rxn} + q_{soln} + q_{calorimeter}$$
or
$$q_{rxn} = -q_{soln} + - q_{calorimeter}$$

$\Delta H = q_{rxn}$ / moles of limiting reagent Mg for part I and MgO part II

YOU WILL CALCULATE q_{rxn}

THEN SOLVE FOR q_{soln} by using the graphs of your reactions to obtain your final temperatures and ΔT

So what are these heats? What units do they have? We may not know what the q rxn is (that's what we want to calculate) but we can find a general form for a heat change when it results in a change in temperature. A heat that arises from a temperature change is equal to the quantity of material times the change in temperature times its specific heat (s). We will use the specific heat of water for all calculations because we are working with aqueous solutions.

Thus for the heat terms given above we can rewrite the solution in water and calorimeter terms as follows:
$q_{soln} = m\,s\,\Delta T$ and $q_{cal} = m\,s\,\Delta T$ (We will use specific heat of water for all reactions in the experiment) where the mass of material is m, s is the specific heat of the material and ΔT is the temperature change is degrees Celsius. For our calculations we will assume that the heat liberated by the reaction is equal to the heat gained by the mass of water (HCl solution) in the calorimeter + the heat gained by the calorimeter. The s of water is well known and has a value of 4.184 J/g-°C; the m-s term for the calorimeter will be estimated to be approximately 10 J/°C. If we can determine a temperature change (ΔT), then we can assign a heat change to the water and the calorimeter. Once that is accomplished we can determine the heat that was liberated in the reaction. In this case remember that the difference between heat generated in a reaction and the enthalpy of reaction is the correction for the number of moles of the limiting reagent.

With the calculation of heat changes possible from the equations above the last obstacle to overcome is the calculation of the temperature change for the reaction. That seems easy enough since there must be a temperature when you start the reaction and there must be a temperature when you stop the reaction. Can't you just subtract the two values and arrive at a T value? Well, no you can't and the reason why is that the chemicals may react at different rates (is this true in the reactions you have observed? A reaction that gives off heat slowly will take a while to reach its maximum temperature.

Also although we have assumed that our calorimeters are well insulated they are not perfect and will "leak" heat to the surroundings. We can use this last piece of information to our advantage, however, and use what is called the "decay curve" to provide us with an excellent value for what temperature change would be if the chemical reaction had been instantaneous and liberated the heat within a very short period (a few seconds).

If we wait until the reaction is complete and graph how quickly the calorimeter loses heat after the reaction we can then determine what the temperature would have been if the reaction had occurred at time of mixing with full liberation of the heat of reaction. In other words, if we know the rate that the temperature of the calorimeter decreases after the reaction is complete, we can <u>extrapolate</u> (plot a line which defines existing data to a point we have no information about) to the time of mixing to determine what the T would have been if the reaction had been instantaneous. This is easier if the reaction is fast than if the reaction is slow. The figures below give examples of the three types of temperature profiles possible (Temperature on Y axis, Time on x axis; arrows represent when reaction starts; double arrows represent ΔT):

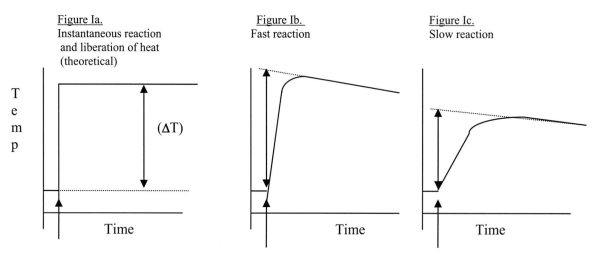

Figure Ia.
Instantaneous reaction
and liberation of heat
(theoretical)

Figure Ib.
Fast reaction

Figure Ic.
Slow reaction

Since we live in the real world we will **not** have data that looks like figure Ia (that is what we are trying to model) but we will have data that will look like Figure Ib or Ic depending on the rate of reaction. After the maximum temperature of the reaction has been obtained we will record the decay curve and extrapolate to the time of mixing. From this we will obtain our ΔT values for the reaction. Substitution into the equations above will provide heat change values q water and q cal. From this you can calculate q rxn and ΔH rxn. Finally use enthalphy of formation data to compare with the values you obtain and use Hess's Law (also using the (ΔH_f value for water) to determine the value for the heat of reaction for the conversion of magnesium metal to magnesium oxide (the reaction run in Expt 3). Compare the value you obtain to that from heat of formation data to assess the accuracy of your experiment.

<u>Reading Assignments:</u>
You are required to complete the following reading assignments as part this laboratory session: General Chemistry Atoms First, *2ⁿᵈ Ed.,* McMurry/Fay, Chapter 8 Thermochemistry: Chemical Energy.

<u>Grading:</u> Grading will be as described in the general laboratory handout.

Procedure:

1. General setup, use of the iBook/LabPro system and reaction of Mg with HCl.

1.1 Set up the calorimeter apparatus as shown at the instructor's desk. You will need two styrofoam coffee cups nested inside one another, a ring stand and support ring to keep the cups upright, a 100 mL graduated cylinder and the iBook/LabPro system including a temperature probe and USB cable. Set up the iBook as far from the coffee cups as possible to avoid damaging the computer.

1.2 Transfer 100 mL of 3.0 M HCl to the calorimeter and insert the temperature probe through the lid of the calorimeter. Do not remove the waxy coating on the temperature probe. This covering is to protect the probe from the HCl solution.

Set Up for Data Collection Using Vernier Temperature Probe

1.3 Turn on the iBook and select Student from the log-on window. The password is kimberly unless noted otherwise by your instructor. Connect the power supply to the LabPro, connect the temperature probe in Channel 1 of the LabPro and connect the USB cable to both the LabPro and the iBook. The USB port is on the right side of the LabPro and the left side of the iBook. The power supply to the LabPro is on the bottom left side of the unit.

1.4 On the iBook look to the bottom of the screen and move the cursor onto the pink square which contains a caliper symbol (this looks like a Greek letter pi with an elongated line on the top). Double click on this icon to open the LoggerPro application on the computer. You should see a screen which says LoggerPro at the top of the screen, a data table on the left and an empty graph window on the right. On the line that says LoggerPro is a series of pull down menus (File, Edit, Experiment, etc.).

1.5 You will now run a short temperature vs. time acquisition using the LabPro system to determine the room temperature and to test the connections of the LabPro/iBook system. Move the cursor to the Experiment pull down menu and select Data Collection. Change the Length window to 60 and set the sampling speed to the center of the bar to give 1 sample/s (and 1 s/sample). Select Done from the bottom of the window. Go back to the Experiment pull down menu and select Start Collection. The data table should fill with data and the graph should show a relatively straight horizontal line across the window. The temperature should be shown actively above the data table. Select the Analyze pull down menu and select Examine. Move the the line across the data set and estimate an average value for the temperature recorded by the probe over this 60 second time interval. Record this temperature in the data table as Room Temperature.

1.6   Using weighing paper determine the mass on the analytical balance to within 0.1 mg of approximately 0.25 g of magnesium ribbon. **Remember to obtain the mass on the analytical balance to 4 significant figures. Data from the top load balance will result in point deductions for the laboratory.**

1.7   Select Experiment from the pull down menu and from that menu select Data Collection. Change the Length to 16 minutes (you must change seconds to minutes by using the downward arrow). Change the Sampling Speed to 6 samples per minute (0.16666 min/sample) which should show Samples to be collected: 97. If you do not have these values on the Collection Setup screen contact the instructor. Once you confirm these values select "DONE" at the bottom right of the window.

1.8   Select the Experiment pull down menu and select Start Collection. Using the time values in the data table (or using a stop watch) record the temperature values every 0.50 minutes (30 seconds). After 30 seconds add the magnesium to the calorimeter and cover the calorimeter. For the first 2-3 minutes stir the calorimeter frequently (every 10 seconds) by removing the cups from the support ring and swirl the liquid inside to make sure all of the Mg has been combined with the HCl solution. Be sure the temperature probe is in the solution (at the bottom of the calorimeter). Keep recording the data for the full 16 minutes and stir the calorimeter occasionally (every 1-2 minutes after the temperature maximum has been reached) in order to ensure mixing. Try to avoid removing the cover during the run since that may allow some of the heat generated by the reaction to escape. If the computer screen goes dark touch the space bar. You should move the bar at the right of the data table to the bottom to have the data table display the current values.

1.9   Once the data has been collected (both manually and by the computer) select the Analyze pull down menu and select Examine from this menu. Move the cursor to the very first part of the data before the magnesium has been added.

1.10   Select the Analyze and then Examine from the pull down menu. Move the cursor to the highest point on the graph (highest temperature) and drag the cursor to form a darker blue window that starts at the highest point on the graph and ends with the last time point (16 minutes). Select Analyze and Linear Fit from the pull down menu. Record the slope and intercept of this line and move the box that displays these values to a portion of the graph that does not contain data. Select the File pull down menu and the Print Graph option. The number of copies should equal the number of members in your group. Select Print from the Print window and get the hard copies from the printer in the lab. Record the average value from the data table from the first three data points (just before the Mg is added) and record this value as the initial temperature. Solve the linear equation for t = 0.500 minutes (provided you added the Mg at 0.50 min) and enter this value in your data table as the Final Temperature (linear regression).

1.11   Repeat 1.6-1.10 for a second trial using magnesium. You may not need to record the temperature vs. time data as you did for the first data set. When you start the data collection save the data when the computer gives that option. Be sure to record the initial temperature and the equation for the linear fit in the data table provided.

2. Reaction of Magnesium Oxide with HCl.

2.1   Repeat 1.6-1.9 substituting approximately 0.5 grams of MgO for the magnesium metal. Remember to record the mass of MgO on the analytical balance to within 0.1 mg! Perform two determinations for the reaction of MgO with HCl and remember to stir frequently to ensure good mixing. In these reactions you may find that there is no significant temperature decrease. You should still determine the linear fit line (linear least squares fit or linear regression line) using the maximum temperature and the time points after the maximum. In these trials (Trial 3 and 4) the slope will probably be close to zero.

3. Calculation of the Heat of Reaction

3.1   Convert the number of grams of magnesium and magnesium oxide used in this experiment to moles and enter these values into the table provided.

3.2   Confirm that you have a value for the initial temperature from the first three data points before Mg was added.

3.3   Graph the first data set as a temperature (y-axis) vs. time (x-axis) graph. It is preferable to have time (x-axis) as the longer dimension on the graph. Draw the best straight line which represents the data from where the temperature starts to decrease to the last data point observed. Extrapolate (extend) this line to the time where you added the Mg metal and determine this temperature as T_f. Record this value in the data sheet provided.

3.4   Confirm that you have a T_f value for all of your four trials by solving the linear fit line described in 1.9. You should also have a T_f value from your manual graph as described in 3.3.

3.5   You should now have the following data for each reaction: T_o, T_f, mass of Mg or MgO and the moles of Mg or MgO. <u>We will assume that the heat properties and density of the HCl solution in the calorimeter have the are the same as pure water. Thus for calculation purposes 100 mL of HCl = 100 mL of pure water.</u> Calculate the heat given off by the reaction using the s of water to be 4.184 J/g-°C, the density of water as 1.00 g/mL, and the m s of the calorimeter as being 10 J/°C. You should have a heat term for the water as well as the calorimeter as follows:

q_{soln} = m s ΔT = (mass of water used) x 4.184 J/g-°C x (T_f - T_o)
q_{cal} = m s ΔT = 10 J/°C x (T_f - T_o)

Determine the values for q_{water} and q_{cal} and enter this information into the table provided.

The heat absorbed by the water and the calorimeter plus the heat given off by the reaction is equal to zero:
q_{total} = 0 = q_{rxn} + q_{soln} + $q_{calorimeter}$

This equation can be solved for the heat given off by the reaction:
q_{rxn} = -q_{soln} + - $q_{calorimeter}$

Calculate the value for q reaction and enter it into the table provided. The enthalpy change for the reaction is the heat given off divided by the number of moles of limiting reagent. Divide the q_{rxn} by the moles of material used and enter this value as the ΔH_{rxn} in the table.

3.6   You should have two values for the enthalpy of reaction as described in Trial 1 (manual graph and from the linear curve fit). Determine the percent error between each of these Trial 1 values and your Trial 2 value.

$$\% \text{ error} = \frac{\text{Trial 1 value} - \text{Trial 2 value}}{\text{Trial 2 value}} \times 100\%$$

3.7   Using the average calculated values for the enthalpy of reaction for each of the reactions (use the value for Trial 1 with the smaller % error as described in 3.6) studied and the accepted value for the reaction between H2 and O2 to form water given below, determine the enthalpy of reaction for the formation of **one mole** of MgO from Mg and oxygen (Hess's Law). Determine this same value from heat of formation data and record this value as the theoretical value. Calculate the % error between your expermental value and the theoretical value from the equation below. Be sure to use units for all of your calculations!

Mg reaction: $Mg_{(s)} + 2HCl_{(aq)} \longrightarrow MgCl_{2(aq)} + H_{2(g)}$ ΔH = expt avg value

MgO reaction: $MgO_{(s)} + 2HCl_{(aq)} \longrightarrow MgCl_{2(aq)} + H_2O_{(l)}$ ΔH = expt avg value

Water reaction: $H_2 + 1/2\, O_2 \longrightarrow H_2O_{(l)}$ ΔH = -286 kJ

$$\% \text{ error} = \frac{\Delta H_{\text{theory}} - \Delta H_{\text{expt}}}{\Delta H_{\text{theory}}} \times 100\%$$

Results:

Room Temperature					
Reactant	Mg	Mg	Mg	MgO	MgO
Trial Number	1	2	3 (optional)	3	4 (optional)
Mass of reactant					
Moles of reactant					
Initial Temperature (T_o)					
Linear Curve fit slope (m)					
Linear Curve fit intercept (b)					
(T_f) (temp) at time = 0.50 min ($Y=T_f$) from equation of line					
$\Delta T = T_f - T_o$					
q_{soln}					
q_{cal}					
q_{rxn}					
ΔH_{rxn}					

Hess's Law determination of ΔH for   Calculation of % Error:

$$Mg + 1/2 O_2 \longrightarrow MgO:$$

Time vs. Temperature Data (manual data collection----MAY BE OPTIONAL)

Trial Number			(cont.)	
Time (min)	Temperature		Time	Temperature
0				

Temperature (y-axis) vs. Time (x-axis) Graph (optional if Vernier software used and graphs saved)

Questions:

1.   One of the assumptions is that the heat generated by the reaction is absorbed by pure water. Calculate the mass of HCl, the mass of Mg in your first trial and the mass of water (assume density = 1.0 g/mL for the HCl solution). Determine the percentage of water by mass in the system. Does this support the assumption that the heat is absorbed by water? Explain you answer.

2. The calculations of the enthalpy of reaction assume that Mg and MgO are the limiting reagents. Prove that this is indeed the case for both the Mg and the MgO experiments.

3. Magnesium oxide reacts slowly with water to form magnesium hydroxide. What would be the effect on the observed value for ΔHrxn for the reaction of MgO if 10% of the MgO had reacted with water to form $Mg(OH)_2$? Would the value be higher or lower? (Hint: you need to determine the ΔH value for the reaction of $Mg(OH)_2$ with HCl from heat of formation data.)

4. Determine the ΔH of reaction for the following reactions using enthalpy of formation data from the textbook (the $ΔH°_f$ of Mg_3N_2 (s) is -461.2 kJ/mol):

    a) $Mg_3N_{2\,(s)} + 6H_2O_{(l)} \longrightarrow 3Mg(OH)_{2(s)} + 2NH_{3(g)}$

    b) $Mg(OH)_{2(s)} \longrightarrow MgO_{(s)} + H_2O_{(l)}$

SCI124 Principles of Chemistry II
Laboratory 2: Solids and Solutions

Purpose:

This laboratory is concerned with the physical properties of solids and solutions. Crystalline solids exist in a lattice of repeating structural units. The student will gain greater appreciation of how atoms are spatially arranged in a solid. The solubility of an ionic compound will be evaluated as a function of temperature. Phase changes associated with super-cooled liquids, supersaturated solutions and sublimation will also be presented.

Introduction:

Everywhere we look we see matter as pure elements, compounds and mixtures. Most of the things that we observe as physical materials are either solids or liquids. Many of the solids we see have a complex structure that is polymeric based from either natural (wood is an example) or synthetic (plastics are an example) sources. Simpler crystalline materials are also observed such as the sugar that we put in our coffee or the salt in the food we eat. We also observe liquid matter normally as water or as solutions of things in water (coffee, tea and soda are examples). It is known that these materials are all comprised of atoms, but how are these atoms arranged? This laboratory session should provide some insight into how atoms and molecules are arranged into the materials that we come into contact with daily.

Crystalline solids exist in an extended network of repeating structural units called a crystalline lattice. This extended network provides additional stability to the solid (related to the lattice energy) and the repeating unit is called a unit cell. We will examine the arrangement of some simple unit cells by making models on a scale we can easily observe. In addition we will calculate the free volume in these structures and calculate the density of some pure metals using these geometric measurements.

Solutions are homogeneous mixtures of one or more compounds in a liquid called the solvent. In our examinations the solvent used will be water and we will determine the solubility of potassium nitrate in water at various temperatures and generate a solubility curve. When determining the solubility of a compound in a solvent, we will be looking for the point at which the solution exists at its maximum concentration that is called the saturation point. We will also examine systems where a compound exists as a liquid *below* its normal melting point and a solution contains more solute than present in a saturated solution. These systems (supersaturated solutions and super-cooled liquids) are said to be thermodynamically unstable and if given enough time will form a solid but require a seed crystal or shock to initiate crystal formation. Finally a demonstration of sublimation will be observed whereby iodine will be converted from a solid to a gas (sublimation) and then back to a solid (deposition).

Reading Assignments:

You are required to complete the following reading assignments as part of this laboratory session: General Chemistry Atoms First, *2ⁿᵈ Ed.*, McMurry/Fay, Chapter 10 Liquids, Solids, and Phase Changes, Sections 10.8-10.9. Chapter 11 Solutions and Their Properties Section 11.4 Some Factors That Affect Solubility

Grading:

Grading will be as described in the general laboratory handout. Remember to have your raw data signed before you leave the laboratory if required.

Procedure:

1. Models of Crystalline Unit Cells

1.1 From the Styrofoam balls available in the laboratory take a package of the full and partial balls to prepare models. Record the radius of the balls used. It may be helpful to refer to Figure 11.22

1.2 Using the Styrofoam balls and toothpicks provided prepare a unit cell for the simple cubic, body-centered cubic and face-centered cubic (also called cubic closest-packed; ccp) crystal lattice unit cells.

1.3 Measure the dimensions (unit cell length, l) of each unit cell with a ruler and include this data in the table provided. Also calculate the ratio of the cell length (all are cubes) to the atomic radius (l/r ratio). Draw what you have made to the best of your ability including a representation of the unit cell dimensions. Determine the number of atoms present in each unit cell and enter this data into the table provided.

1.5 Place your finished face centered cubic unit cell at the instructor's desk along with other student's unit cells to create a portion of a face- centered cubic lattice structure.

1.6 Calculate the following and enter into the table provided:
1.6.1 The dimensions (geometric unit cell length, l) of the unit cell as a function of radius by using geometry (as opposed to direct measurement of the model). Note: For the body and face-centered cubic you will need to apply the Pythagorean theorem ($c^2 = a^2 + b^2$ for a right triangle).
1.6.2 The volume of the 'atoms' used to prepare the models ($4/3\pi r^3$)
1.6.3 The total volume of the unit cell occupied by the atoms (number of atoms x volume of each atom) and the total volume of the cell (volume of the cube, l^3).
1.6.4 The percentage of space occupied by atoms (volume of space occupied/volume of unit cell x 100%) and the percentage of free space for each unit cell (100% - Pct. filled).

1.6 Using the information on crystal structures from Figure 1 below and the following atomic radii, calculate the density that you would expect for the metals Cu (r = 128 pm), Fe (r = 124 pm), and Pb (r = 175 pm). Compare the values you obtained with the reported values for density listed in the table. Note: You will need to calculate the mass (using Avogadro's number) and volume (using the atomic radius and the volume of a sphere) of each metal atom.

Figure 1 Simple Cubic, Body-Centered Cubic and Face-Centered Cubic Unit Cells

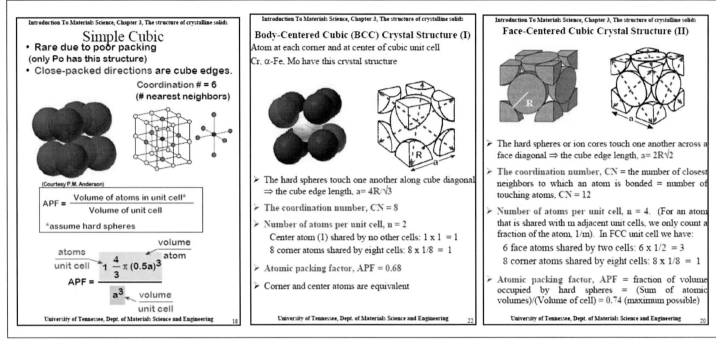

Figure 1 is courtesy of University of Tennessee, Knoxville Dept. of Materials Science and Engineering

2. Determination of the Solubility Curve for Potassium Nitrate.

2.1   Using a heating block set near 90°C (or a water bath) heat the samples described below to up to 90 degrees C. (Note that higher concentrations will require higher temperatures to dissolve.) The instructor may request that you substitute a thermocouple temperature probe for the thermometer.

2.2   Add the following amounts of potassium nitrate to four separate test tubes: 2, 4, 6, and 8 grams. Record the weight to within at least 0.01 grams.

2.3   Pipette 5.00 mL of water into each test tube.

2.4   Heat the test tubes until the samples completely dissolve using a glass stir rod to gently mix the contents. Do not remove the glass stir rod from the mixture until it is completely dissolved.

2.5   Move the test tube containing the KNO_3 solution to a test tube rack and quickly place the thermometer into the KNO_3 solution. If the sample crystallizes upon the addition of the thermometer then heat the sample and the thermometer together in the heat block until re-dissolved.

2.6   Slowly raise and lower the thermometer in the test tube and record the temperature at which a crystal is first observed. Record this temperature as accurately (within 0.1 degree) as possible. This is the temperature of a saturated KNO_3 solution.

2.7   Repeat with the other concentrations of KNO_3 recording the temperature for each tube (Note: The temperature required to completely dissolve the sample will be lower at lower concentrations).

2.8   Graph the solubility of KNO_3 in g/100 mL as a function of temperature on the graph paper provided.

2.9   Dispose of the KNO_3 solutions in the container provided in the laboratory.

3. Super-cooled Liquids and Supersaturated Solutions (Acetic Acid and Sodium Acetate)

3.1   Remove a bottle of pure acetic acid which has been cooled from the instructor's desk. Record the temperature of the ice/water bath that the acetic acid was held in and compare to the melting point of acetic acid. Handle the bottle carefully and observe if there is any solid present in the solution. Dry off the thermometer and use the thermometer to gently stir the solution and record any observations.

3.2   Take one of the flasks of concentrated sodium acetate solution which is at or below room temperature. Make sure that there are no crystals evident in the solution used (if there are then use another flask of sodium acetate which does not have crystals).

3.3   Observe carefully the beaker containing the sodium acetate solution which has been cooled and confirm that there is no solid present.

   3.4 Add a spatula of sodium acetate crystals to the sodium acetate solution and record the results

4. Sublimation (Iodine)

4.1   Observe the demonstration at the instructor's desk where iodine is being sublimed. What do you observe about the solid in the bottom of the beaker and the solid on the watch glass? What about the "air" in the beaker. Record your observations.

Results:

1. Models of Crystalline Unit Cells

	Simple Cubic SC	Body Centered Cubic BCC	Face Centered Cubic FCC (CCP)
Atomic Radius (radius of Styrofoam ball)	r = (include units!)		
Volume of one atom (one ball, $4/3\pi r^3$)	V = (include units!)		
Measured unit cell length (l) [1.3]			
No. Atoms per cell [1.3]			
Ratio of l/r [1.3]			
Geometric unit cell length (l) [1.5a]			
Volume of atoms in cell [1.5c]			
Total Volume of Unit Cell [1.5c]			
Percent vol filled [1.5d]			
Percent free space [1.5d]			

Drawings of Unit Cells (Include cell dimensions) [1.3]		
Simple Cubic	Body Centered Cubic	Face Centered Cubic
SC	BCC	FCC (CCP)

	Copper	Iron	Lead
Crystal Form	FCC	BCC	FCC
Radius (pm)	**128**	**124**	**175**
No. atoms per cell (See Figure 1)			
Dim. of unit cell			
Volume of unit cell			
Mass of one Atom			
Total mass of atoms in one unit cell			
Mass of atoms per unit cell vol (calculated d)			
Reported density (g/cm^3)	**8.9**	**7.9**	**11.3**

2. Determination of the Solubility Curve for Potassium Nitrate

Approx g KNO3 per 5 mL water	Mass of KNO3 (grams)	Volume Water (mL)	Saturation Temperature (°C)	Calc. g KNO3 per 100 mL
2		5.00 mL		
4		5.00 mL		
6		5.00 mL		
8		5.00 mL		

Graph of Results (Solubility/100 mL water vs. Temperature)

Grams of KNO3 per 100 grams of water

Temperature (°C)

3. Supersaturated Solutions (Acetic Acid and Sodium Acetate)

<u>Acetic Acid</u>
Temperature of water bath _____°C, mp of acetic acid = 16.6 °C

Observations before shaking the bottle:

Observations after shaking the bottle:

<u>Sodium Acetate</u>

Observations before adding seed crystal:

Observations after adding seed crystal:

4. Sublimation (Iodine)

Record your observations about the demonstration at the instructor's desk.

Questions:

1. How do the densities of the metals calculated in section 1.6 correlate with the reported values for these metals? Could this method be used for determining the density of mercury as well? Briefly explain your answer. (Hint: what do you know about mercury as an element?)

2. Water is a unique material in that the density of the solid is lower than the density of the liquid (which is why ice forms at the top of a pond and why ice floats in our drinks). If the density for ice at 0 °C is 0.917 g/mL and the density for water at 0 °C is 0.999 g/mL, what is the calculated free space (as %) for each of these materials? You will need to estimate the volume of water as the sum of 2 H atoms and 1 O atom with radii of 37 and 66 pm respectively. Note that you will also have to assume a quantity of water to perform this exercise.

3.   From the information in the phase diagram of carbon dioxide (Figure 10.29) would the method used to sublime iodine at atmospheric pressure also work to sublime and collect dry ice (solid carbon dioxide)? Explain your answer briefly. (Hint: At what temperatures would dry ice exist at 1 atmosphere? Remember you used this material last semester in a gas laboratory exercise).

4.   Would the method used to determine the solubility curve for potassium nitrate work for sodium acetate as well? Why or why not?

SCI 124 Principles of Chemistry II
Laboratory 3: Colligative Properties of Solutions: Molecular Weight by Freezing Point Depression

Purpose:

This laboratory examines the freezing point depression of a solution containing various amounts of solute. The molecular weight of the solute is obtained from the magnitude of the freezing point depression. A discussion of colligative properties and the calculation of molecular weight by accurate determination of the freezing point depression are emphasized.

General:

Colligative properties are those properties that result in some measurable change in a physical property of a solution but do not depend upon the chemical composition (identity) of the solute. In other words they are _general_ changes that result upon the addition of any solute. Two common colligative properties are freezing point depression and boiling point elevation. In our day-to-day lives we see the utility of these in the salting of roads to prevent icing in cold weather (salt dissolves in the water to give a solution which melts below that of pure water) and the addition of anti-freeze (ethylene glycol) to car cooling systems which provide both freezing point depression (your radiator will not freeze even at temperatures much lower than $0^{o}C$) and boiling point elevation (your radiator will not 'boil over' even when the coolant temperature exceeds $100^{o}C$).

The magnitude of these changes depend upon the number of solute molecules present in the solution (again, not dependent upon the chemical identity of the solute) or to rephrase, on solute concentration. Normally when we quantitatively evaluate colligative properties we find that there is a linear relationship between the concentration in mole fraction or molality and the property examined. In this laboratory you will observe the depression of freezing point on a solution of p-dichlorobenzene (an organic solid) dissolved in cyclohexane (an organic solvent). You will measure the magnitude of the freezing point depression by determining the freezing point of the pure solvent (cyclohexane) and the freezing point of the solution of solute in solvent. The difference between the pure solvent freezing point and the solution freezing point is the freezing point depression (ΔT_f). From a known amount of solute (in grams) and an observed ΔT_f the molecular weight of the solute can be determined. You will determine the molecular weight of p-dichlororbenzene and compare it to the known value for this material.

Reading Assignments:

You are required to complete the following reading assignments as part of this laboratory session:
General Chemistry Atoms First, _2^{nd} Ed.,_ McMurry/Fay, Chapter 11 Solutions and Their Properties. Section 11.7 Boiling Point Elevation and Freezing Point Depression of Solutions.

Grading: Grading will be as described in the general laboratory handout.

Procedure:

General: Cyclohexane is a volatile organic compound so minimize the amount of cyclohexane in the air by keeping bottles or test tubes of this material closed when not actively transferring this material. Any appreciable loss of cyclohexane from the apparatus will affect your results. It is important that all glassware and apparatus used in this experiment be **clean and dry**. This laboratory procedure is written for using a thermocouple attached to a LabPro/iBook system.

1. Freezing Point of Pure Cyclohexane

1.1   Prepare an ice-water bath with at least 30-40% of the volume consisting of ice. This is most easily done using a large (600mL) beaker. Add 20 grams of NaCl to the ice-water bath and stir to dissolve to make it colder.

1.2   Pipet 10.00 mL of cyclohexane into the clean dry test tube provided. Place the stopper, stirrer and thermocouple into the test tube but do not place the test tube in the ice bath at this time. You should secure the test tube using a clamp and an ring stand but keep the test tube above the ice bath.

1.2   Turn on the iBook and select Student from the log-on window. The password is bhcc unless noted otherwise by your instructor. Connect the power supply to the LabPro, connect the temperature probe in Channel 1 of the LabPro and connect the USB cable to both the LabPro and the iBook. The USB port is on the right side of the LabPro and the left side of the iBook. The power supply to the LabPro is on the bottom left side of the unit. Set up the iBook as far from the test tube as possible to avoid damaging the computer.

1.3   On the iBook look to the bottom of the screen and move the cursor onto the pink square which contains a caliper symbol (this looks like a greek letter pi with an elongated line on the top). Double click on this icon to open the LoggerPro application on the computer. You should see a screen which says LoggerPro at the top of the screen, a data table on the left and an empty graph window on the right. On the line that says LoggerPro is a series of pull down menus (File, Edit, Experiment, etc.).

1.4   You will now run a short temperature vs. time acquisition using the LabPro system to determine the room temperature and to test the connections of the LabPro/iBook system. Move the cursor to the Experiment pull down menu and select Data Collection. Change the Length window to 60 and set the sampling speed to the center of the bar to give 1 sample/s (and 1 s/sample). Select Done from the bottom of the window. Go back to the Experiment pull down menu and select Start Collection. The data table should fill with data and the graph should show a relatively straight horizontal line across the window. The temperature should be shown actively above the data table. Select the Analyze pull down menu and select Examine. Confirm that the plot is a straight line (although it may be noisy if the y scale represents a small range of temperature).

1.5   Select Experiment from the pull down menu and from that menu select Data Collection. Change the Length to 15 minutes (you must change seconds to minutes by using the downward arrow). Change the Sampling Speed to 6 samples per minute (0.16666 min/sample) which should show Samples to be collected: 91. If you do not have these values on the Collection Setup screen contact the instructor. Once you confirm these values select "DONE" at the bottom right of the window.

1.6 Select the Experiment pull down menu and select Start Collection. Using the time values in the data table (or using a stop watch) record the temperature values every 0.50 minutes (30 seconds). *Once the data collection has started place the test tube into the ice bath such that the entire volume of cyclohexane is surrounded my ice and water.* Start stirring the test tube continuously with the stirrer throughout the experiment (until solid such that you can no longer stir the sample). Keep recording the data for the full 15 minutes and stir the test tube in order to ensure adequate mixing. If the computer screen goes dark touch the space bar. You should move the blue bar at the right of the data table to the bottom to have the data table display the current values.

1.7 Once the data has been collected (both manually and by the computer) select Analyze from the pull down menu and select Examine from this menu. Move the cursor to form a window that represents the region of temperature that is relatively constant after the initial temperature decrease. (Ask the instructor if uncertain about where to measure the data.) Select Analyze and Statistics from the pull down menu. Write down the average (mean) temperature in this region (in the small box) and obtain a copy of this data (File/Print).

1.8 Remove the test tube from the ice bath and allow to warm back to near room temperature (it should not feel very cold by touch). Repeat this procedure to obtain a second value for the freezing point of cyclohexane. Allow the apparatus to warm to near room temperature and **do not discard the cyclohexane from the apparatus since you will use the same cyclohexane in the next part of the experiment**.

2. Melting point of cyclohexane solutions

2.1 Weigh out 0.10-0.15 grams of p-dichlorobenzene on an analytical balance. Record the mass to within 0.1 mg (**do not use a top loading balance**).

2.2 Add the p-dichlorobenzene to the apparatus, seal with the stopper and stir the solution until the solid completely dissolves. Be certain that all of the solid is dissolved and no residual solid remains on the walls of the apparatus.

2.3 Repeat the data collection described in part 1 (1.5-1.10) for the cyclohexane solution. In this case the temperature should decrease as the solution freezes. Once the data collection is complete remove the sample from the ice bath but do not discard the solution.

2.4 Select Analyze and Examine from the pull down menu and move the cursor such that the temperature data includes only the drop in temperature from room temperature to before the sample starts to freeze. This should be close to a straight line in this region. Select Analyze and Linear Fit from the pull down menu. Record the best fit line equation in place in the data table indicated. Again select Analyze and Examine but now use the cursor to define the slower temperature decrease that occurs as the sample is freezing (again close to a straight line). Select Analyze and Linear Fit from the pull down menu and record the best fit line equation for this data in the data table. Obtain a hard copy of your graph from the computer (File/Print).

2.5 Remove the apparatus from the ice bath and allow it to warm to near room temperature but do not discard the cyclohexane solution.

2.6 Weigh a second sample of p-dichlorobenzene on an analytical balance using half as much solid as in 2.1 (mass for sample 2 should be between 0.06-0.10 grams) and again record the exact mass to within 0.1 mg. Record the <u>sum</u> of the two net weights as the total amount of p-dichlorobenzene in the table.

2.7   Repeat 2.2-2.4 by adding this second sample to the apparatus (note that the total amount of p-dichlorobenzene is the sum of the two masses for this part of the experiment).

2.8   Once you have the second solution data (written and hard copy from the computer) discard the cyclohexane solution in the waste container provided and rinse the test tube with acetone and place on the cart.

<u>3 Determination of Freezing points.</u>

A. Pure cyclohexane (1.1-1.8)

3.1   The freezing point of a pure material such as cyclohexane should not change during the freezing process. As heat is removed from the cyclohexane solution it turns to a solid (freezes) until all of the cyclohexane becomes a solid. The heat removed during the freezing process is related to the heat of fusion for cyclohexane. Once all of the cyclohexane has solidified then the temperature will decrease until it matches the temperature of the ice bath (0 degrees C). Thus a temperature-vs.-time graph of pure cyclohexane in the first two trials should be similar to figure Ia (below).

3.2   In order to determine the freezing point of pure cyclohexane you must find the portion of the curve during which the temperature does not substantially change. This constant temperature is the freezing point of pure cyclohexane. If manually graphing the data, draw a line through the constant temperature portion of the data and extend the line to the Y-axis in order to determine the freezing point of cyclohexane. If determining by computer the average which was recorded from the graph (and in the box on the graph) will represent the freezing point. Erratic changes in data or portions of the curve which show a substantial (greater than 0.3 degree) increase in temperature normally indicate poor mixing during the experiment. Data that shows this type of behavior is less reliable for obvious reasons.

3.3   Determine the freezing point of pure cyclohexane by obtaining the average values for trial 1 and trial 2. Record this data in the table provided.

3.4   The temperature vs. time plot of a solution (not a pure material since it contains solute) is **not constant** as it freezes. This is because the concentration of solute in the solution changes during the freezing process. A typical curve for a cyclohexane solution is given in Figure Ib.

3.5   You can obtain the freezing point of a cyclohexane solution by placing the two equations found the linear fit of the data equal to one another. When you solve for the time from this equality you determine the time at which the two lines intersect. Solving either equation using this value for time will give you a value for the freezing point. You should use this method to determine the freezing point of each solution. In addition your instructor may ask you to manually determine the freezing point graphically.

3.6   In order to graphically determine the freezing point of a solution you need to determine two lines from the data. The first line (L1) describes the cooling of the solution before any solid is formed.

The second line (L2) describes the cooling of the solution during the freezing process. Graph the data you obtained for the cyclohexane with p-dichlorobenzene added (Trial 3, 2.1-2.5). Draw the best fit straight line (L1) for the initial portion of the curve from approximately room temperature to 8 degrees C. Extend this line (called extrapolation) to a temperature equal to the final temperature. Draw a second line (L2) which describes the portion of the curve during the freezing process (approximately 6 to 3 degrees C) and extrapolate (extend) this line until it intersects with L1. The intersection of L1 and L2 represents the freezing point of the solution. Draw a short "work line" which shows where the intersection of L1 and L2 falls on the Y-axis. Record this temperature as the freezing point for the cyclohexane solution (Trial 3). Again, erratic data or increasing temperatures are indicative of inefficient or incomplete mixing during the experiment.

3.7   Repeat 3.5 or 3.6 for Trial 4 (with the additional quantity of p-dichlorobenzene added) as you did for Trial 3. Record your data in the table provided.

Pure Cyclohexane (In both cases the arrow marks the freezing point) and Cyclohexane soln.

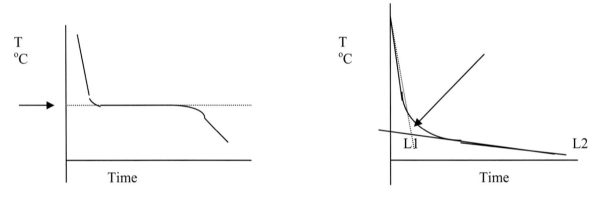

4. Calculation of Molecular Weight of p-Dichlorobenzene

Note: You will have freezing point values from both the computer and by graphical analysis. Use the data from the computer method for the calculations below unless your instructor indicates otherwise.

4.1   Calculate the mass of cyclohexane used in this experiment in kg given that cyclohexane has a density of 0.779 g/mL. Enter this value in the table provided.

4.2   Calculate the true value of the molecular weight of p-dichlorobenzene given the this material has a molecular formula of $C_6H_4Cl_2$. Enter this value in the table provided.

4.3   Calculate the freezing point depression (ΔT_f) for Trial 3 and Trial 4 by subtracting the T_f obtained from the average value for the freezing point of pure cyclohexane. Enter this value into the table provided.

4.4   The freezing point depression (ΔT_f) is related to the concentration of solute in molality (c_m, moles of solute per kg of solvent) according the the following equation:

$\Delta T_f = K_f(c_m)$ where the constant $K_f = 20.5\ ^oC/m$ for cyclohexane Rearrange this equation and using the values obtained for ΔT_f in Trial 3 and Trial 4 calculate the molality of the p-dichlorobenzene/cyclohexane solutions for each of these trials. Enter these values in the table provided.

4.5   From the definition of molality and the known amount of cyclohexane that was used in the experiment, calculate the number of moles of p-dichlorobenzene from the calculated solution molality for each trial. **(DO <u>NOT</u> DETERMINE THE NUMBER OF MOLES OF SOLUTE BY DIVIDING THE MASS OF P-DICHLOROBENZENE ADDED BY ITS TRUE MOLECULAR WEIGHT.)**

4.6   Calculate the molecular weight of p-dichlorobenzene from the mass of solid used in trial three and the number of moles calculated to be present in trial three from above (4.5). Perform the same calculation for trial 4 (remember to use the sum of the masses for the total amount of solute in trial 4). Enter this data into the table provided.

4.7   Determine the average value for the molecular weight of p-dichlorobenzene and enter it into the table. Calculate the percent error between the average value of MW obtained and the true value as follows:

% error = <u>Expt. value - True value</u> x 100%
                         True value

Record the percent error value in the table provided.

Results:

1. **Solvent and Solute Amounts**

A. Cyclohexane [4.1]		mL of cyclohexane
Mass used (d=0.779g/mL)		Mass in kg =
B. p-Dichlorobenzene	**Trial 3**	**Trial 4**
Net Mass [2.1, 2.6]		
	Total Mass for Trial 4 [2.6]	

2. **Determination of Freezing Points**

Cyclohexane	Trial 1	Trial 2
T_f by computer		
Average T_f		
Cyclohexane solutions	Trial 3	Trial 4
Linear Fit of Initial Temp Drop (L1) [2.4]		
Linear Fit of Freezing Temp Drop (L2) [2.4]		
Solve for time at intersection of lines (make two lines equal) [3.5]		
T_f for solutions [2.4, 3.5-3.7]		
Graphical Determination (Trial 3 or Trial 4)		
	Trial 3	Trial 4
ΔT_f for cyclohexane solutions [4.3]		
molality (c_m) of cyclohexane solutions [4.4]		
number of moles of solute present [4.5]		
MW of solute [4.6]		
MW of p-dichlorobenzene [4.2, 4.7] $C_6H_4Cl_2$	True =	Avg =
% Error [4.7]		

Raw Data (Temperature vs. Time manual—optional)   A. Pure cyclohexane

Trial 1				Trial 2			
Time	Temp	Time	Temp	Time	Temp	Time	Temp

Raw Data (Temperature vs. Time manual---optional) B. Cyclohexane Solutions

Trial 3					Trial 4			
Time	Temp	Time	Temp		Time	Temp	Time	Temp

Questions:

1.   From your understanding of colligative properties, how would you design an experiment to measure the freezing point of a material which freezes at -2 $^{\circ}$C using the same apparatus used in this laboratory and any chemical reagents commonly found in a chemical laboratory. Ice is the only material you have available to cool the solutions.

2.   If you measured the boiling point of the solution used in Trial 4, would you predict it would have a boiling point equal to, higher than, or lower than the boiling point of pure cyclohexane? Explain your answer.

3.   Calculate the molecular weight of an unknown substance if 0.2064 grams of solid was added to 25.00 mL of cyclohexane and the resultant solution exhibited a freezing point depression of 2.1 degrees C.

4.   What would be the effect on your calculated molecular weight of p-dichlorobenzene if your lab partner left the stopper off of the apparatus while you were weighing the solid p-dichlorobenzene. Explain in detail what would happen if this occurred and how it would affect your result.

THIS GRAPH PAPER IS SUPPLIED BELOW SHOULD YOUR INSTRUCTOR ASK YOU TO GRAPH
YOUR EXPERIMENTAL RESULTS FROM PART I INSTEAD OF/IN ADDITION TO THE VERNIER
DATA COLLECTION AND GRAPHING

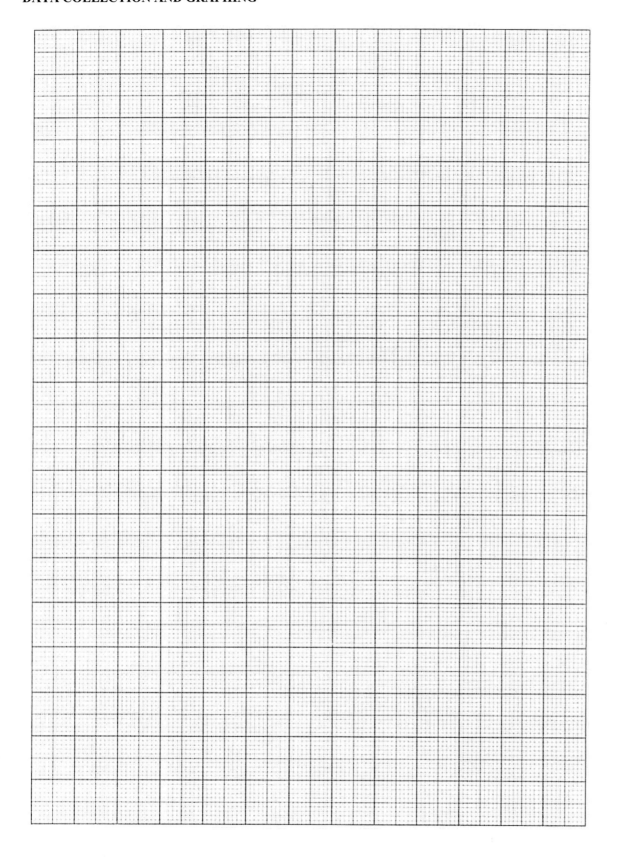

THIS GRAPH PAPER IS SUPPLIED BELOW SHOULD YOUR INSTRUCTOR ASK YOU TO GRAPH YOUR EXPERIMENTAL RESULTS FROM PART II INSTEAD OF/IN ADDITION TO THE VERNIER DATA COLLECTION AND GRAPHING

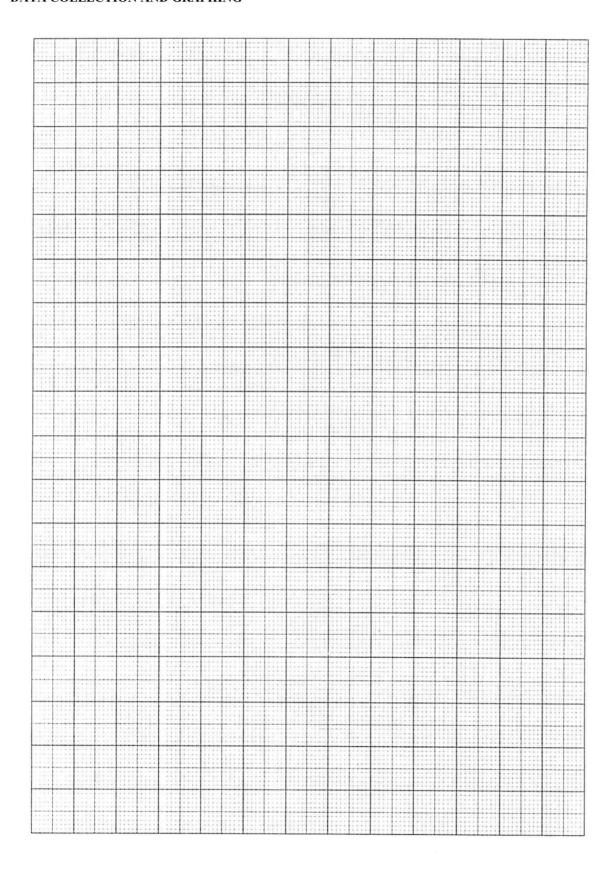

SCI 124 Principles of Chemistry II
Laboratory 4 Separations Lab: Column Chromatography of Grape Dyes and TLC of Curcumin from Tumeric

References: Samuella B. Sigmann and Dale E. Wheeler, *Journal of Chemical Education*, **2004**, *81*, 1475-1478
Andrew M. Anderson, Matthew S. Mitchell, and Ram S. Mohan, *Journal of Chemical Education*, **2000**, *77*, 359-360

Purpose:

Compounds of similar polarity will tend to be soluble in one another. This laboratory examines some of these solubility trends and how methods based on these solubility phenomena can be used to separate one component from a more complex mixture. The use of chromatography to separate mixtures into their component compounds will be examined. This will include both thin layer chromatography (TLC) as well as using a reverse-phase sep-pak C-18 cartridge. The food dyes separated by reverse phase chromatography will be analyzed quantitatively by visible spectroscopy.

Introduction:

The phrase "like dissolves like" is used to describe the observation that compounds of similar polarities will dissolve in one another. Examples of this rule include that ionic compounds such as sodium chloride dissolve in very polar solvents such as water and that cyclohexane (C_6H_{12}, a non-polar organic solvent) and water do not dissolve (are immiscible). This laboratory session will evaluate this solubility behavior for ionic and organic solutes in both polar and non-polar solvents. How solubility behavior changes from acid-base reactions will also be examined. Chromatography is a separation method that is based on the attraction of polar (or non-polar) compounds to one another. We will examine how compounds that are chemically similar can be separated from one another using chromatographic methods.

Chromatography

Chromatography is an extremely useful method for separating materials from natural sources and from synthetic reactions. Chromatography requires the interaction of the material of interest with a solid material (stationary phase) and a moving gas or liquid (mobile phase). If one transfers material from a green leaf by rubbing it onto a piece of filter paper and placing the filter paper in an organic solvent, it is possible to isolate the green chlorophyll as a green streak on the paper once the organic solvent has evaporated. This is an example of paper chromatography since the chlorophyll interacts with both the paper (stationary phase) and the organic solvent (mobile phase). We will be isolating the yellow pigment curcumin from the spice turmeric. In Thin Layer Chromatography (TLC) there is a separation due to the rate at which components move up a plate covered with a solid when a solvent travels over the plate. Substances that have little attraction for the solid (stationary phase) will tend to move faster than materials that have a strong attraction for the solid. We will measure the ratio of the distance traveled by the substance we are looking for to the distance traveled by the solvent (the ratio is called the **R_f value**). In column chromatography the sample is passed through a tube containing the solid material (stationary phase) by using a solvent (mobile phase). The rate at which the sample moves through the tube will depend on how much it is attracted to the solid in the tube. Some components of a mixture will move through quickly since there is little attraction between the component and the solid. In both TLC and column chromatography the rate at which the sample passes through the stationary phase depends on its attraction to the solid relative to its attraction to the mobile phase. As the mobile phase polarity matches more closely the polarity of the stationary phase, the compounds present will move faster (have higher Rf values) through the stationary phase.

For the TLC experiment the stationary phase is silica gel that you can consider finely divided sand (SiO_2). When sand is broken down to the small particles we find in silica gel the surface attracts water and the silica gel particles are covered with a large number of OH groups attached to silicon (silanol groups). As a result in TLC analysis the stationary phase (silica gel) is very polar. Thus non-polar solvents will move compounds more slowly (weak solvents) than polar (strong) solvents. For exactly the same reasons using the same solvent a less polar compound in the mixture will move faster than more polar compounds. In the Sep-pak analysis of the dyes used in grape soda the solid which is the stationary phase is said to be a reversed phase support. This means that the normal polarity of the silica gel has been reversed and the OH groups on the surface have been chemically changed to long chain hydrocarbons (18 carbons). As a result the stationary phase in the sep-pak is non-polar. This means that water is a weak solvent and the Rf values will increase (samples will move faster) as the polarity of the mobile phase decreases. We will observe how mobile phase polarity affects both of these separations. For TLC you will measure Rf values after increasing the polarity of the solvent mixture; for the sep pak experiment you will isolate one dye with a higher water concentration than required to isolate the second dye (the red and blue dyes in grape soda).

Thus normal phase chromatography using silica gel in the TLC of curcumin in tumeric (a spice) has a polar solid (stationary phase). If analytes (the compounds you are trying to measure) differ in polarity then the more polar compound will have the lower Rf value (will move slower up the plate). For a given compound you are observing (analyte) the Rf will increase as the polarity more closely matches the polarity of the solid. In this case increasing the solvent polarity (mobile phase) will result in higher Rf values. In reverse phase chromatography the polarity of the solid is low (non-polar), less polar compounds will have lower Rf values and the Rf increases (moves faster through the sep pak column) as the solvent polarity decreases.

Thin Layer Chromatography Analysis of Curcumin in Tumeric

Curcumin is the yellow coloring found in the spice tumeric. Tumeric comes from a ginger-like plant in Asia called *Curcuma longa* but you may be more familiar with this spice as it is a component in curry powder. The major pigment in tumeric is a compound called curcumin with the structure given below. The way in which this structure is shown is common for complicated organic molecules and these line drawings indicate that there is a carbon at every point a line segment changes angles and hydrogens are not shown to make it simpler. As a result curcumin has a molecular formula of $C_{21}H_{20}O_6$ and has a molecular weight of 368.4 grams/mole (or amu). One thing that you may notice from the structure is that it has a large number of double bonds that alternate with single bonds. Organic chemists would say that this molecule is highly conjugated and this is a general structural feature of organic compounds that have an observable color (absorb visible light).

curcumin, $C_{21}H_{20}O_6$

Thin layer chromatography is a technique in which a sheet of plastic is coated with silica gel on one side (the non-shiny or dull side). You add the sample to be analyzed as a small dot at the bottom of the plate, normally 0.5 to 1.0 cm from the bottom of the short dimension of the plate. The sample is added to the plate as a solution using a fine capillary tube to transfer the sample. The solvent evaporates off of the plate leaving the spot of colored material on the plate. The plate is then placed into a glass jar that has solvent in the bottom. The spot should be above the level of the solvent in the jar. The jar is sealed loosely with a cap and solvent moves up the plate in the same way that water moves up a paper towel placed into a drop or pool of water (capillary action). As the solvent moves up the plate the sample can either stick to the plate (stationary phase) or move with the solvent (mobile phase). When the solvent is approximately 1 cm from the top of the plate the plate is removed from the jar and the solvent level is marked with a pencil. You can measure the Rf by dividing the distance traveled by the top darkest spot (the curcumin) by the distance traveled by the solvent.

Analysis of the Dye Content in Grape Soda

You may not be surprised to find that grape soda has both artificial coloring and flavoring. The purple color comes from the addition of colorants (dyes) approved in the US by the Food and Drug Administration (FDA) and are given FD&C (Food, Drug and Cosmetics) numbers. In grape soda the two dyes most commonly used are FD&C Red #40 which is also called allura red and FD&C Blue #1 also called neptune blue. The structures of these dyes are given below and as you see for curcumin these are also relatively large organic molecules with a large number of double and single bonds that alternate (are highly conjugated). The red dye (red #40) absorbs at 505 nm and the blue dye (blue #1) absorbs at 620 nm. The absorption of the dye can be related to the concentration by the Beer-Lambert law used in previous labs. The absorbance (A) is related to the concentration (c) directly. The relationship is $A = \varepsilon b c$ where A=absorbance, ε = molar absorptivity (sometimes called the extinction coefficient), b=path length in cm and c= molar concentration. In all of our measurements the path length is fixed at the standard value of 1.00 cm. As a result concentration is related to the absorbance from the molar absorptivity (ε). A table of these values is given below that includes the formula, molecular weight and molar absorptivity for both of the dyes in grape soda. The concentration of dye in the samples that you isolate will be equal to the absorbance divided by the molar absorptivity.

Dye compound	Wavelength Nm	Formula	MW (amu or g/mol)	molar absoptivity (ε) ($M^{-1}cm^{-1}$)
FD&C Red #40	505 nm	$C_{18}H_{14}N_2Na_2O_8S_2$	496.43	25,900
FD&C Blue #1	620 nm	$C_{37}H_{43}N_2Na_2O_9S_3$	792.84	130.000

The grape soda (5.0 mL) is forced through a sep pak cartridge which contains C-18 modified (reverse phase) silica gel which is non-polar. As a result the most "organic like" compounds in the mixture which includes the dyes will be adsorbed onto the solid in the cartridge while sugar, water and other more polar compounds will not stick to the C-18 solid phase. Once the dyes have been transferred to the sep pak the dyes can be removed by flowing solvents through the cartridge. If you start with water or highly polar solvents (have a high water percentage compared to the organic solvent) then only the least easily retained (least polar) compounds will stay on the column. As you increase the organic content of the solvent and make it less polar you will remove more strongly held compounds. The cartridge will be treated with 5% and then 20% isopropanol (an organic solvent) in water. This should cleanly separate the more polar (Red dye #40) from the less polar (Blue #1) dye.

Structures of dyes used in this experiment

FD&C #40, $C_{18}H_{14}N_2Na_2O_8S_2$

FD&C #1 Blue
$C_{37}H_{43}N_2Na_2O_9S_3$

Grading:
Grading will be as described in the general laboratory handout.

Reading Assignments:

General Chemistry Atoms First, *2nd Ed.,* McMurry/Fay, Chapter 11 Solutions and Their Properties. Section 11.1 Solutions and Section 11.2 Energy Changes and the Solution Process. HANDOUTS: Chromatograpy: TLC and Hydrophobic Interaction Columns (Reverse Phase).

Procedure:

General: You should keep a 600 mL beaker at your workstation to use as a satellite waste receptacle. Use this waste receptacle for aqueous (water based) solutions only. Dispose of the organic (cyclohexane) solutions in the separate container provided using the ethanol wash bottle to rinse off the organic solutions. Cyclohexane evaporates readily so be sure you are observing a solubility result not simply the evaporation of solvent. You may need to transfer the materials into a disposable pipette or mix the wells with a toothpick or small plastic stirrer to clearly see the solubility/miscibility results.

1. Qualitative examination of solubilities.

1.1   Two solids and two solvents of very different polarity will be used in this portion of the experiment. The non-polar solid is stearic acid, the polar (ionic) solid is sodium chloride, the non-polar solvent is cyclohexane and the polar solvent is water.

1.2   Using a porcelain spot plate add as small amount as you can readily observe of solid stearic acid in two of the wells adjacent to one another. To the first sample of stearic acid add 25 drops of water and to the second sample add 25 drops of cyclohexane. Mix the materials and record your observations. If it appears that there was partial dissolution of the solid in either case add another 5-10 drops of the solvent to be sure of your observations.

1.3   Repeat 1.2 substituting sodium chloride for the stearic acid.

1.4   Transfer a small amount of stearic acid to one of the wells on the plate. Add 30 drops of cyclohexane and mix then add 25 drops of water. Collect this mixture in a disposable plastic transfer pipette and invert the pipette such that the liquids are in the bulb of the pipette (SA+ CH +water). THIS IS LIKE A MINI-SEPERATORY FUNNEL.

1.5   Shake the pipette to ensure the liquids are well mixed and record what this sample looks like -- is there one layer (one phase) or two layers? If there are two layers which one contains cyclohexane and which contains the water? How could you test this?

1.6   Repeat 1.4-1.5 using 0.01M NaOH in place of the water (SA+CH+NaOH).

1.7   Using the pipette from 1.6 containing the NaOH solution with the liquid in the bulb of the inverted pipette, curve the tip of the pipette into a well containing 0.1 M HCl and add 25 drops of the acid solution to the pipette and mix with the existing solution. Shake the pipette and record your observations (add HCl to the SA + cyclohexane + NaOH solution).

1.8   Repeat 1.4 substitution sodium chloride for the stearic acid and adding the water prior to the cyclohexane (NaCl + water + cyclohexane).

1.9   Repeat 1.8 using 0.01 M NaOH in place of the water (NaCl + NaOH + cyclohexane).

1.10   Repeat 1.7 for this sample (add HCl to the solution in 1.9).

2.   TLC Analysis of curcumin in turmeric

2.1   A sample of turmeric has been extracted into an organic solvent prior to the laboratory session. You will use this material for your analysis of curcumin.

2.2   Using a TLC applicator (a fine capillary with a 5-20 µL volume) spot a sample of the turmeric extract on the analytical TLC plate approximately 1 cm from the bottom edge of the plate. You should be able to see a colored spot where you have applied the sample. If not then repeat application method until the spot is visible. Mark with a pencil (not pen!) on the edge of the plate where the samples have been applied.

2.3   Pour a small amount of the developing solvent (5% methanol [CH_3OH] in dichloromethane [CH_2Cl_2]; the mobile phase) into the jar such that the solvent completely but just barely covers the bottom of the jar. Place the plate in the jar with the spot edge down (lower) making sure that the solvent level is below the spot on the plate. Develop the TLC plate until the solvent is within 1 cm of the top of the plate. Remove the plate from the chamber and mark on the edge of the plate how high the solvent has traveled on the plate.

2.4   Make a sketch of the plate and record the distance traveled by the solvent starting at where you spotted the sample. The curcumin spot should be the top and darkest spot on the plate. Record the distance traveled by curcumin and all other spots below the curcumin on the plate.

2.5   Record the Rf values for all of the spots on the plate by dividing the distance traveled by the spot by the distance traveled by the solvent. Make a note as to which of the spots was the darkest and any characteristic color or appearance of each of the spots on the plate.

2.6   Run a new TLC plate in the same way that you did the first plate but add one or two drops of methanol to the solvent mixture (increasing the polarity of the mobile phase) before running the plate. Record the Rf values and your observations as in 2.4 and 2.5 above.

3. Separation and Analysis of Dyes in Grape Soda

3.1   Draw 10.0 ml of 70.0% IPA solution into the syringe. Connect the syringe to the Sep-Pak. Push the IPA solution in the syringe through the cartridge slowly. Discard the solution in the waste container or a beaker at your work space.

3.2   Repeat 3.1 using 10.0 ml of water in place of the 70% IPA solution. This pretreats the cartridge and replaces the relatively non-polar IPA solution with water (polar). The cartridge is now ready to use for the separation.

3.3   Set up a test tube rack with 15 test tubes capable of containing at least 10 mL each. Draw 5.0 ml of Grape Soda into the syringe; slowly force it through the Sep-Pak. Collect the solution that comes through the Sep-Pak (the eluent or first fraction) into the first test tube. The first fraction should be colorless and the Sep-Pak should now be purple (the color of Grape soda).

3.4   Draw up 5.0 mL of 5% IPA solution into the syringe and slowly force it through the Sep-Pak. Collect the eluent into the second test tube to give the second fraction. Fraction 2 should have pink color as Red Dye #40 is removed from the Sep-Pak.

3.5   Repeat step 3.4 three more times placing the eluent into separate test tubes to give fractions 3, 4 and 5. The color of fraction 5 should be lighter than fraction 4 since most of the red dye should be removed once 20 mL of 5% IPA solution has been passed through the Sep-Pak.

3.6   Draw 2.5 ml of 5% IPA into the syringe and push it through the Sep-Pak. Collect the eluent into a separate test tube as fraction 6. Repeat this four more times to give fractions 7-10 which should each have 2.5 mL of solution. Fraction 6 may still have a light pink color but some of the fractions between 6 and 10 should have no observable pink or blue color. You may also see some blue starting to elute in fraction 10. The Sep-Pak should still have a blue color.

3.7   Draw 5.0 ml of 20% IPA into the syringe and slowly force it through the Sep-Pak (as you did in 3.4) collecting the eluent in the next test tube to give fraction 11. Fraction 11 should have a definite blue color.

3.8   Repeat step 3.7 until the eluent has no observable blue color. This should require less than 20.0 mL of the 20% IPA solution (no more than 4 test tubes using the 20% solution).

3.9   Repeat 3.1 and 3.2 to clean the Sep-Pak and return the Sep-Pak to the beaker provided on the cart or at the instructor's desk.

3.10   Combine all of the tubes which have a red or pink color and measure the total volume using a graduated cylinder. Record the volume of the combined fractions.

3.11   Set up the spectrophotometer to measure at 505 nm and set at 0.0%T with no cuvette in the instrument and set at 100.0 %T using a cuvette containing distilled water. Measure the absorbance of this solution at 505 nm and record the absorbance value.

3.12   Repeat 3.10 above for all of the tubes which have a blue color. As for the other combined fractions measure the total volume of the combined fractions and record this value.

3.13   Repeat 3.11 above for the combined blue fractions but perform the measurements at 620 nm. Be sure to adjust the 0%T and 100%T values at the new wavelength and check to be sure that the filter is set correctly for 620 nm (the lever on the front left of most Spec-20s).

4. Calculations for Absorbance Data

4.1   From the absorbance value at 505 nm for the combined red fractions calculate the concentration in molarity of Red Dye #40 using the molar absorptivity value provided ($\varepsilon = 29,500$ M^{-1}cm^{-1}). Note that the path length (b) of the cuvettes used is the usual standard value of 1.00 cm.

4.2   Repeat 3.1 for the blue fractions at 620 nm using the molar absorptivity of the blue dye at 620 nm ($\varepsilon = 130,000$ M^{-1}cm^{-1})

4.3   From the concentration of the red dye and the ***total volume of the combined fractions*** (not 5.00 mL!), calculate the number of moles of FD&C Red Dye #40 in your 5.0 mL sample. Do the same calculation for the blue dye (FD&C Blue Dye #1)

4.4   From the number of moles calculated in 4.3, determine the mass of each dye using the molecular weights provided (Red Dye MW is 496.43 g/mole; Blue Dye MW is 792.84 g/mole) present in the 5.0 mL sample that you analyzed. Determine the mass of each dye that would then be present in a 355 mL can of soda and in a 2.0 L bottle.

Results:

1. **Qualitative examination of solubility behavior.**

Ref	Solute	Solvent	Observations
1.2	Stearic Acid	Water	
1.2	Stearic Acid	Cyclohexane	
1.3	Sodium Chloride	Water	
1.3	Sodium Chloride	Cyclohexane	
1.4-5	Stearic Acid	Cyclohexane + Water	
1.6	Stearic Acid	Cyclohexane + NaOH solution	
1.7	Stearic Acid	HCl to prev. soln.	
1.8	Sodium Chloride	Cyclohexane + Water	
1.9	Sodium Chloride	Cyclohexane + NaOH solution	
1.10	Sodium Chloride	HCl to prev. soln.	

2.   **TLC Analysis of curcumin in turmeric**

Sketch the appearance of the TLC plates and show the values traveled by the solvent and each spot in the mixture. If curcumin is the highest Rf dark spot, identify which of the spots would correspond to curcumin. Below each plate record the Rf values for each spot and any comments you have about the appearance of each spot in the mixture.

5% Methanol in Dichloromethane   > 5% Methanol in Dichloromethane

SHOW RF CALCULATIONS HERE:

HOW DOES THE INCREASED POLARITY OF THE SOLVENT AFFECT THE RF FOR CURCUMIN?

3.   **Separation and Analysis of Dyes in Grape Soda**

Observations on the colors of the fractions:

	FD&C Red #40 (Allura Red) $C_{18}H_{14}N_2Na_2O_8S_2$ MW =496.43 ε = 25,900 (at 505 nm)	FD&C Blue #1 (Neptune Blue) $C_{37}H_{43}N_2Na_2O_9S_3$ MW =792.84 ε = 130,000 (at 620 nm)
Fraction Numbers Combined		
Total Volume (mL)		
Aborbance		
Concentration (M) [4.1 or 4.2]		
moles of dye in 5.0 mL [4.3]		
mg of dye in 5.0 mL [4.4]		
mg of dye in one can (355 mL) [4.4]		
mg of dye in one 2.0L bottle [4.4]		

Questions:

1. Does the solubility of stearic acid and sodium chloride follow the "like dissolves like" pattern? Explain if this rule is followed or not.

2. What reaction occurs when sodium hydroxide is added to the stearic acid in cyclohexane and water? How does this change the solubility of cyclohexane in water? Does HCl reverse this reaction and if so why? Does sodium chloride have the same type of behavior as stearic acid or is it different? Explain why these two solutes are the same or different.

3. How would the Rf values change if you used pure dichloromethane (no methanol) as the mobile phase for the TLC separation of curcumin? What if you used pure methanol (no dichloromethane)?

4. Would the separation of the grape soda dyes work equally well if you reversed the order of IPA concentrations and used 20% IPA first to remove the blue dye and then switched to 5% IPA to remove the red dye? Explain what you would expect to happen if you reversed the order of solvents in this separation.

5. If you had a dye with a molecular weight of 574.76 amu which had a molar absorptivity of 82,000 at 550 nm, what would be the expected value of absorbance at that wavelength if the dye was present at a concentration of 1.0 mg/L? What color would you expect for this dye?

6. The structure of β-carotene which is a yellow dye found in vegetables that is also important in vision is shown below. From the structure and formula would you expect this compound to have a higher or lower Rf than curcumin? Explain your prediction AND BE SPECIFIC ABOUT WHAT SOLVENT SYSTEM YOU ARE USING (POLAR OR NON-POLAR).

SCI 124 Principles of Chemistry II

Laboratory 5: Chemical Kinetics: The Iodine Clock Reaction

Purpose:

This laboratory examines the rate of chemical reactions in solution. For a series of reactions run a room temperature, the concentration of reactants will be varied such that the order of reaction can be determined for two reactants. The effect of temperature and catalysts on the reaction rate will also be examined. This series of experiments will emphasis the determination of kinetic parameters from the initial rate method including the determination of an experimental rate law, values for rate constants and the order of reaction (overall and reactant specific).

General:

Our discussions of chemical reactions to date in Principles of Chemistry have included how to balance equations, how to measure the heat given off or generated, and how to calculate the amount of products or reactants from the number of moles of other materials present. We will now consider how fast reactants are transformed into products which are the field of chemistry known as chemical kinetics. In order to study the rate at which a product is produced from one or more reactants we need to understand and control a number of experimental variables. The following is a list of things that can affect the rate at which a product can be formed in a chemical reaction:

1) Solution concentration -- more concentrated reactants form products faster than more dilute reactants.

2) Temperature -- Reactions are almost always faster at higher temperatures and in order to understand reaction rates we must carefully control the temperature of our experiments.

3) Catalysis -- Some materials can act to increase the reaction rate without being part of the net observed equation. These materials are called catalysts and may have a large effect on the reaction rate even when present in small (sometimes undetectable) quantities.

4) Surface area of a solid reactant or catalyst -- The more surface area (more finely divided solid) the faster the reaction if there is a solid-liquid or solid-gas phase interface as a component of the chemical reaction.

From this list we will primarily evaluate changes in concentration and temperature. We will also determine the effect of a catalyst on our reaction. Since the reaction we will study involves the reaction of aqueous solutions with no solids present as reactants, we will not be concerned with surface area (#4 above).

In order to understand the rate at which reactants combine to form products we have to have a clear understanding of what reaction we are studying and how we can measure the reaction rate. Sometimes this is straightforward such as the decomposition of hydrogen peroxide given below:

$$H_2O_{2(aq)} \longrightarrow H_2O_{(l)} + O_{2(g)}$$

In this reaction there is one gaseous substance and in this reaction it is a product. If you measure the gas volume of oxygen produced (in other words connect a closed vessel containing the reaction to a gas collection apparatus identical to what you used to measure the molecular weight of butane) as a function of time you would be able to clearly measure the increase in the number of moles of oxygen.

From this previous example the rate of product formation can be measured directly. By varying the concentration of hydrogen peroxide you can determine the rate law and the rate constant of this reaction.

The reaction that we will study in this laboratory session is called an iodine clock reaction. It is called a clock reaction because the completion of the reaction (similar to a titration endpoint) is signaled by the sudden appearance of the dark blue starch-iodine complex at a fixed time for the same concentration of reactants. In order to obtain this dramatic color change and relate that to the chemical reaction rate requires us to perform two reactions at the same time. The first reaction (primary reaction) is the reaction of kinetic interest which is the reaction of iodide ion with persulfate ion to form iodine and sulfate ion. Is this reaction a precipitation, acid-base, or oxidation-reduction reaction? How can you tell? If it is an acid-base reaction then identify the acid and the base; if it is an oxidation-reduction reaction identify what is being oxidized and what is being reduced. The molecular and net ionic equations are as follows:

Molecular: $(NH_4)_2S_2O_8 + 2KI \longrightarrow I_2 + (NH_4)_2SO_4 + K_2SO_4$

Net Ionic: $S_2O_8^{2-} + 2I^- \longrightarrow I_2 + 2SO_4^{2-}$
$\quad\quad$ (persulfate ion) (iodide ion) $\quad\quad$ (iodine) (sulfate ion)

If we want to measure the kinetics then we want to experimentally determine the rate at which persulfate ion or iodide ion is consumed or the rate at which iodine or sulfate ion is produced. If you run this reaction in the laboratory you observe that upon reaction the solution turns yellow-orange which is an indication that iodine is formed. One experimental possibility to determine the reaction rate would be to determine how dark the iodine color is as a function of time with an instrument (spectrophotometer). Since you already know that starch will form a blue color with iodine you could perform the same reaction in the presence of starch. If you did so you would observe that a dark blue color forms almost upon mixing. To determine the rate this way would be difficult because it is hard to tell the difference between dark blue, darker blue and even darker blue! Thus the direct method that could be employed in the oxygen production reaction of hydrogen peroxide cannot readily be used in this case.

In order to provide a reaction that can be more easily observed we will run a second reaction (probe reaction) at the same time the first reaction is proceeding. This second reaction will quickly consume all of the iodine produced in the primary reaction until the probe reaction uses up all of its reagents. When the probe reaction stops it will be <u>only</u> because all of the reactants have been used up and iodine will form from the primary reaction. If we add starch into the system the end effect will be the fast dramatic color change from colorless to dark blue (iodine starch complex). This rapid color change is why it is referred to as an iodine clock reaction. The probe reaction that we will use is (precipitation, acid-base or oxidation-reduction reaction?):

Molecular: $2Na_2S_2O_3 + I_2$ (from primary reaction) $\longrightarrow$ $2NaI + Na_2S_4O_6$

Net Ionic: $2S_2O_3^{2-} + I_2 \longrightarrow 2I^- + S_4O_6^{2-}$
$\quad\quad$ (thiosulfate ion) (iodine) $\quad\quad$ (iodide ion) (tetrathionate ion)

Because the probe reaction is fast (faster than the primary reaction) and goes to completion it gives you an important piece of information: <u>The number of moles of iodine (and for a fixed volume the concentration in molarity) produced by the primary reaction in a measured amount of time.</u> You know the time because you can measure when the blue color appears. You know the amount of iodine that was produced because it is related to the amount of thiosulfate you started with (all experiments will use the same amount of thiosulfate). Thus the appearance of the dark blue starch-iodine complex means that all of the thiosulfate has been consumed and the starch-iodine complex is formed from additional iodine arising from the primary reaction. Furthermore you know the exact time (which you measure) that all of the thiosulfate has been consumed and therefore you know how much iodine was generated at that point in time. You have information that can relate the molar concentration change of a product as a function of time and from this you can determine other kinetic parameters.

Let's look at some conditions that we could recreate in the lab and see what we would expect to observe for each case. You should try to write down the concentrations of iodine asked for in the last two columns of the table below. We are concerned only with three ions present in the solution: thiosulfate ion ($S_2O_3^{2-}$), persulfate ion ($S_2O_8^{2-}$) iodide ion (I^-). From the concentrations of these three ions predict the amount of iodine consumed.

Persulfate Ion Conc (M)	Iodide Ion Conc. (M)	Thiosulfate Ion Conc. (M)	Time	Iodine Conc Present at t (M)	Iodine Conc Consumed (M)
1.0	2.0	0.0	Enough to fully react	1.0	0.0
1.0	2.0	1.0	Enough to fully react	0.50	0.50
1.0	2.0	2.0	Enough to fully react	0.0	1.0
1.0	2.0	0.50	Enough to fully react	0.75	0.25
1.0	2.0	0.10	Enough to fully react	0.95	0.050
1.0	2.0	0.10	First formation of Blue	>0M	0.050
2.0	2.0	0.10	First formation of Blue	>0M	0.050
1.0	4.0	0.10	First formation of Blue	>0M	0.050
2.0	4.0	0.10	First formation of Blue	>0M	0.050

If we use amounts of thiosulfate less than the amount of iodine that can be produced and if we always use the same amount of thiosulfate then the blue color will appear when a known quantity of iodine (based on the amount of thiosulfate initially present in the solution) has been consumed by all of the available thiosulfate. The concentration of iodine produced is always one half of the initial concentration of thiosulfate both from the reaction and the table above.

$$\text{Rate} = \frac{\Delta[I_2]}{\Delta t} = \frac{0.5\ [Na_2S_2O_3]_{initial}}{\text{Time to Blue Color}}$$

The rate of a chemical reaction is the concentration increase of product (or decrease of reactant) over a given time. As discussed above you now know the relationship between concentration of iodine produced at the appearance of the blue color (from the known concentration of thiosulfate used but note that you will have to calculate the initial concentration of all reactants) and the time that the blue color appears. This rate in M/sec (mol/L-sec) is the experimentally measurable chemical rate for the primary reaction. Now that we know how to measure the rate of a reaction what can we do to more fully understand the kinetics of this process? There will be three major experimental goals in this lab which can be obtained from the experimental determination of the reaction rate:

- Determination of the general and specific (experimental) rate law including the order of reaction with respect to both iodide ion and persulfate ion, and the calculation of the rate constant for this reaction (with appropriate units). This is the primary focus of the laboratory. You should be familiar with how to determine the rate law from initial rate data as described in Example 13.3 in the Chang text.

- Determination of the temperature sensitivity to rate of reaction by measuring the rate at different temperatures. The calculation of the activation energy using the experimental rate constants will be asked in one of the questions in the laboratory.

- The qualitative analysis of the effect of a catalyst (copper ion in this case) in order to determine if it accelerates the reaction and if so by how much.

You will determine the experimental rate law by changing the concentration of the reactants (iodide and persulfate ions) and observing the reaction rate and how it is affected by these concentration changes. In kinetic runs where the iodide or persulfate ions are below the maximum amounts (4.0 mL) there is a non-reactive material (KCl for KI and $(NH_4)_2SO_4$ for $(NH_4)_2S_2O_8$) which is added to keep the experiments at the same ionic concentration. Once you have determined the order of the reaction you can use the rates obtained to calculate the rate constant for the reaction at room temperature. You can then accomplish #2 above by performing the same reaction at elevated temperature. You will be asked to use the rate law determined for this reaction to predict the time for a different set of concentrations of persulfate and iodide (experiment 4) and using the Arrhenius equation you will predict the rate constant for the experiment run at a different temperature. Finally you will observe any measurable changes in reaction rate by observing the addition of a catalyst.

 Reading Assignments:

You are required to complete the following reading assignment as part of this laboratory session: General Chemistry Atoms First, *2nd Ed.,* McMurry/Fay, Chapter 12 Sections 12.1-12.7 and 12.14.

Grading:

Grading will be as described in the general laboratory handout.

Procedure:

General: You should keep a 600 mL beaker at your work station to use as a satellite waste receptacle. Use this waste receptacle for the kinetic runs that you have finished. Use a toothpick or some other object to fully mix solutions after all materials have been combined (normally after the addition of persulfate)

1. Qualitative examination of primary and probe reactions.

1.1   On a shallow well plate combine the following materials and record your observations:
    1. 5 drops of DI water + 2 drops of persulfate and 2 drops of KI.
    2. Same as above but with 1 drop of starch added.
    3. Add thiosulfate drop wise to each of the solutions above until the well is colorless.
    4. 3 drops of thiosulfate, 1 drop of starch and three drops each of KI and persulfate [3/1/3/3].
    5. 2 drops of thiosulfate; starch, KI, and persulfate as above [2/1/3/3].
    6. 1 drop of thiosulfate; starch, KI, and persulfate as above [1/1/3/3].

1.2   Set up a miniature of your kinetic runs by using the same amount of thiosulfate and changing the persulfate and KI amounts. Again record your observations:
    Set up 4 wells each containing 1 drop of thiosulfate and 1 drop of starch.
    Add KI 2/4/2/2 drops across the wells and mix the wells.
    Add persulfate (quickly!) 2/2/4/2 drops across the wells and mix.
    Question: Do all of the wells turn color at the same time? Which was the fastest? The slowest?

2. Kinetic Rate Data at Room Temperature.

General: This experiment can be performed at a standard scale using test tubes or on a micro scale using microcentrifuge (Eppendorf) tubes. Your instructor should indicate which procedure you should follow. For the standard scale you will need at least 3 large test tubes with rubber stoppers and will transfer the solutions using graduated pipettes. For the microscale you will use an automatic transfer (Eppendorf) pipette and microcentrifuge tubes. Each kinetic run (experiment with the same concentration values) will be performed twice (in duplicate).

2.1   Record the room temperature in the space provided.

2.2   Using the table below combine the materials together that correspond the amounts in the table. Combine all the materials except the persulfate. You can arrange the data as you please running identical trials together, two independent sets of four, etc.

PICK ONE METHOD EITHER STANDARD OR MICROSCALE—YOU DO NOT NEED TO DO BOTH

Standard procedure (test tube)

Expt. No.	0.20 M KI (mL)	0.20 M KCl (mL)	0.0150 M $Na_2S_2O_3$ (mL)	Starch (mL)	0.10 M $(NH_4)_2SO_4$ (mL)	0.10M $(NH_4)_2S_2O_8$ (mL)
1	2.0	2.0	1.0	1.0	2.0	2.0
2	2.0	1.0	1.0	1.0	0.0	4.0
3	4.0	0.0	1.0	1.0	2.0	2.0
4	1.5	2.5	1.0	1.0	1.0	3.0

Microscale (eppendorf tube) procedure

Expt. No.	0.20 M KI (μL)	0.20 M KCl (μL)	0.0150 M $Na_2S_2O_3$ (μL)	Starch (μL)	0.10 M $(NH_4)_2SO_4$ (μL)	0.10M $(NH_4)_2S_2O_8$ (μL)
1	200	200	100	100	200	200
2	200	200	100	100	0	400
3	400	0	100	100	200	200
4	150	250	100	100	100	300

2.3   For a given kinetic run (remember you have combined all of the materials together except for the persulfate) add the required amount of ammonium persulfate [$(NH_4)_2S_2O_8$], start recording the time using a stopwatch or watch, and stopper and mix the solution thoroughly as quickly after the addition as possible. Mix the contents of the solution occasionally as you observe the mixture. The blue-black color of iodine-starch will develop quickly BE READY FOR IT! and record the time at which you see the color change.

2.4   Perform each of the experiments above in duplicate (8 kinetic trials at room temperature). If the values for duplicate runs do not match within reasonable error (<15% DIFFERENCE) contact the instructor.

3. Kinetic Rate Data at Other Temperatures. USE ONLY EXPERIMENT 1 CONCENTRATIONS AND VOLUMES FOR DIFFERENT TEMPERATURES.

3.1   Prepare a water bath that is between 10 and 18 degrees and combine all of the reactants as you would **for experiment 1ONLY**. In a separate test tube place 5 mL of ammonium persulfate solution for the standard procedure or 500 µL in a separate microcentrifuge tube for the microscale. Place both of these test tubes (or microcentrifuge tubes) in the water bath for 5 minutes and add more cool water if necessary to maintain the temperature within 3 degrees. YOU WANT THE SOLUTIONS TO BE THE CORRECT TEMPERATURE BEFORE THE EXPERIMENT STARTS SO IT IS MOST ACCURATE. After the 5 minutes, pipette the amount of cooled persulfate solution described in the table above (2.2) into the other test tube and mix thoroughly. Keep the test tube (or microcentrifuge tube) in the water bath and shake occasionally. Record the time observed when the iodine starch complex is formed. Try to keep your water bath at as constant a temperature as you can.

3.2   Repeat 3.1 using a hot water bath about 35 degrees C (or a hot water bath may be provided). Record the time at which the iodine-starch complex is observed.

4. Qualitative Evaluation of a Catalyst.

4.1   Set up a kinetic run the **same as experiment 1** in the table above (2.2). Before adding the persulfate add one drop of $CuSO_4$ solution and mix the contents of the test tube. Add the required amount of ammonium persulfate and record the time at which the blue iodine-starch complex is observed.

<u>5. Calculations:</u>

5.1 Calculate the change in iodine concentration during the observed time interval (reaction rate). Remember the iodine concentration change is equal to 0.5 $[Na_2S_2O_3]_{initial}$ and therefore the rate must be this value divided by the Δt observed to form the blue color. The initial concentration of sodium thiosulfate will be the molarity of the sodium thiosulfate concentration transferred times the volume used divided by the total volume (as described in 5.2 below for KI)

$$Rate = \frac{\Delta[I_2]}{\Delta t} = \frac{0.5\ [Na_2S_2O_3]_{initial}}{Time\ to\ Blue\ Color}$$

5.2 Calculate the initial concentrations of both KI and ammonium persulfate $((NH_4)_2S_2O_8)$ for each kinetic run and enter these values in the table provided. Remember that the initial concentration will be the concentration of the reagent used times the volume used divided by the total volume. For example for Experiment 1: [KI] initial = (2.0 mL * 0.20M)/13.0 mL or 200µL * 0.20M/1300 µL)

5.3 Calculate the reaction rate for each of the kinetic runs by dividing the $\Delta[I_2]$ by the time observed for the formation of the blue color. For reactions where you had more than one kinetic run, use the average of the two trials.

5.4 Using the values of the rates for experiments 1-3 at room temperature determine the specific form of the rate law. The general form of the rate law is: Rate = k $[KI]^p$ $[(NH_4)_2S_2O_8]^q$. Using the initial rates method find two experiments where only one of the reactants changes concentration. If you choose the rate where the concentration of the KI is 2X of Expt. 1 and divide the rate of that experiment with that of experiment 1, this should equal 2^p. A similar analysis can be done to solve for q. Once you know p and q write them as the exponents to provide the specific form of the rate law.

5.5 Since you now know the specific form of the rate law and the initial concentrations of both KI and ammonium persulfate, you can calculate values for the rate constant k by rearranging the equation to solve for k. Remember to use correct units!

5.6 Using the specific form of the rate law, an average value for the rate constant from test tubes 1-3 and the initial concentrations of KI and persulfate from experiment 4, calculate the expected rate for experiment 4. Compare this value with the rate you measured experimentally for experiment 4 (calculated in 5.3).

5.7 Using the specific form of the rate law and the initial concentrations from experiment 1, calculate the rate constants for the kinetic runs at higher temperature, lower temperature and in the presence of copper sulfate.

Results:

1. Qualitative examination of primary and probe reactions

Mixture described in 1.1	Observations
1. water/$S_2O_8^{2-}$/KI [5/2/2]	
2. water//KI/starch [5/2/2/1]	
3a. 1 above after addition of $S_2O_3^{2-}$	
3b. 2 above after addition of $S_2O_3^{2-}$	
4. $S_2O_3^{2-}$/starch/ $S_2O_8^{2-}$/KI [3/1/3/3]	
5. $S_2O_3^{2-}$/starch/ $S_2O_8^{2-}$/KI [2/1/3/3]	
6. $S_2O_3^{2-}$/starch/ $S_2O_8^{2-}$/KI [1/1/3/3]	

Small scale kinetics runs [1.2]

All wells have 1 drop of starch and 1 drop of $Na_2 S_2O_3$				
KI (drops)	2	4	2	2
$S_2O_8^{2-}$/ (drops)	2	2	4	2
Time				

2. Kinetic Rate Data at Room Temperature.

Room Temperature (^{o}C) [2.1]	$\Delta[I_2] = 0.5\ [Na_2S_2O_3]_{initial}$ [5.1]

Experiment Number	Time Elapsed	
	Trial 1	Trial 2
1.		
2.		
3.		
4.		
	Temperature (^{o}C)	**Time Elapsed**
1. [3.1] **(Low Temp)**		
1. [3.2] **(High Temp)**		
1. [4.1] **(RT, Cu^{2+} cat)**		

5. Determination of initial rates

Expt. No.	[KI]$_{initial}$ (M) [5.2]	[S$_2$O$_8^{2-}$]$_{initial}$ (M) [5.2]	Average Δt (sec) [5.3]	Rate (M/s) [5.3]
1.				
2.				
3.				
4.				
1. (low T)				
1. (high T)				
1. (RT, cat)				

Determine the specific form of the rate law [5.4]

Rate (M/s) = k [KI]^p[(NH$_4$)$_2$S$_2$O$_8$]^q (solve for p and q)

Value of p		Value of q	
Specific form of Rate Law:	**Rate (M/s) = k [KI] [(NH$_4$)$_2$S$_2$O$_8$]**		

Expt. No.	Value of k (using correct units!)
1.	
2.	
3.	
4.	
1. (low T)	
1. (high T)	
1. (RT, cat)	

Prediction of Rate for Experiment 4

Calculation of the rate in Experiment 4 [5.6]	
Actual value of rate for Experiment 4	

Questions:

1. How does the rate differ in the presence of a catalyst? Do you think that the catalyst would be more effective at higher concentrations? Do you have any suggestions as to how this catalyst might work (hint: what are the charges on the reactants of interest and the catalyst)? Briefly explain your answer.

2. Determine the percentage difference between the experimental and the predicted value for the rate in experiment 4 [5.6].

    % difference = $\frac{\text{rate predicted - rate observed}}{\text{rate predicted}}$ x 100%

    How well do you think you could predict rates at other concentrations? If you have what you consider a large % difference how would you run the experiment to make it better?

3.　　Use the Arrhenius Equation (McMurry/Fay Section 12.10) to calculate the activation energy Ea for this reaction using the Experiment No. 1 data at room temperature and the higher temperature (near 35 degrees C [3.2]; remember to use K for temperature). Using the activation energy (Ea) at room temperature and the rate constant at room temperature, calculate the rate constant at the lower temperature. How well do the two rate constants compare (the rate constant calculated from the Ea and the rate constant from data obtained in [3.1])? NOTE: REMEMBER YOU ARE COMPARING 2 RATE CONSTANTS WHICH CHANGES WITH TEMPERATURE SO YOU MAY HAV A LARGE % DIFFERENCE

SCI 124 Principles of Chemistry II
Laboratory 6: Determination of an Equilibrium Constant

Purpose:

This laboratory session examines equilibria in solution and the calculation of an equilibrium constant. A brief discussion and observation of Le Chatelier's Principle and its relationship to dynamic equilibria is presented. The determination of an absorbing species using a spectrophotometer and the construction of a Beer's Law plot are also included in the laboratory session.

General:

You should now have an appreciation of the fact that when chemicals react that they form products at a certain rate from your studies of chemical kinetics. It turns out that there are many chemical reactions that generate products but not in the molar amounts described by the balanced chemical equation. In fact when these reactions come to completion as defined by a state where there is no more observable concentration change, both products and reactants are present in the reaction mixture. Furthermore if you isolated and purified the reaction product(s) and prepared a solution at the same concentration you would have expected for the first reaction to have at completion, you would find that you have that same mixture of reactants and products. It doesn't matter whether you start with reactants or products you will get the same reaction mixture because this is an example of a **reversible reaction or equilibrium reaction**. In viewing the kinetics of these types of reactions what is observed is that the reaction doesn't stop when the final reaction mixture is formed but that the rate at which reactants being converted to products <u>equals</u> the rate at which products are being converted to reactants. This state of equal rates is called a dynamic equilibrium since the reactions have not stopped but the rates are now equal so the mixture will not change in composition over time.

Chemical equilibria (reversible reactions) are described by writing the reactants and products with arrows in both directions to indicate that these reactions can proceed from reactants to products or from products to reactants. If we examine the reaction you will be examining in this laboratory session, the equilibrium reaction is:

$$Fe_3^+{}_{(aq)} + SCN^-{}_{(aq)} \rightleftharpoons Fe(SCN)^{2+}{}_{(aq)}$$

This reaction has been chosen because the reaction to form product and the reverse reaction are fairly fast, it all occurs in solution, and the product has a characteristic deep red color unlike the reactants (iron ion is light yellow and thiocyanate (SCN^-) is colorless). This allows us a means of determining the amount of product under different conditions if we could quantitatively measure the red color in the solution. Remember that color is related to the wavelength of light that we observe and that a colored solution is one that absorbs some wavelengths of light but not others (if it absorbed all wavelengths of visible light it would be black; if it absorbs no wavelengths of visible light we view it as colorless). You will use an instrument called a spectrophotometer which can measure how much light is absorbed at a given wavelength. You will measure the absorbance of the solutions you prepare at 450 nm which happens to be the wavelength at which the product absorbs the strongest. The higher the absorbance value, the more $Fe(SCN)^{2+}$ must be present in the solution. You will change the concentrations of iron ion and thiocyanate ion and determine the amount of product formed from the absorbance values of the solution. The mathematical relationship between the absorbance value and the concentration of $Fe(SCN)^{2+}$ complex is:

$$A_{450nm} = \varepsilon bc = (\varepsilon b)\,[Fe(SCN)^{2+}]_{eq} = (constant)[Fe(SCN)^{2+}]_{eq}$$

(ε = molar absorptivity or extinction coefficient which is related to the substance you are measuring {$Fe(SCN)^{2+}$}, and b is related to the size of the cuvette or test tube used in the instrument)

This relationship is called Beer's Law and you will use this relationship experimentally to determine the concentration of product formed for different concentrations of reactants.

Equilibrium equations have both products and reactants present when in a state of dynamic equilibrium but how much of each? It turns out that the concentration of products and reactants are related by an equilibrium constant which remains constant regardless of concentration values (provided that the temperature and pressure remain constant). The concentrations are raised to the power of the coefficient in the balanced chemical equation. In the equation you are examining all of the coefficients are 1 so the relationship between products and reactants is given below. Note that this expression becomes more complicated when coefficients in the balanced chemical equation vary.

$$K_c = \frac{[Fe(SCN)^{2+}]_{eq}}{[Fe^{3+}]_{eq}[SCN^-]_{eq}}$$

You will determine the equilibrium concentration of the complex by absorption measurements and the equilibrium concentration of the free ions by knowing how much you originally added and how much exists as the complex. From this you will calculate the equilibrium constant for this reaction.

Reading Assignments:

You are required to complete the following reading assignment as part of this laboratory session: General Chemistry Atoms First, *2nd Ed.*, McMurry/Fay, Chapter 13 Chemical Equilibrium: The Extent of Chemical Reactions Sections 13.1-13.4 and 13.8.

Grading: Grading will be as described in the general laboratory handout.

Procedure:

General: You should keep a 600 mL beaker at your work station to use as a satellite waste receptacle. At the end of the laboratory session transfer the contents of your satellite waste container to the labeled container at the instructor's laboratory bench.

1. Qualitative examination of equilibria and Le Chatelier's Principle

1.1   Using a ceramic spot plate or plastic multi-well plate, add the following materials to five of the wells

Well No.	Contents
1.	3 drops of 0.0025 M Fe(NO3)3
2.	3 drops of 0.0025 M KSCN
3.	1 drop 0.0025M Fe(NO3)3 + 1 drop 0.0025M KSCN
4.	2 drops 0.0025M Fe(NO3)3 + 2 drops 0.0025M KSCN
5.	3 drops 0.0025M Fe(NO3)3 + 3 drops 0.0025M KSCN

Mix each of the wells completely. Is there any difference in the color from well 3 to 5? Is that what you expect? There is more material in the well five but is the concentration different or the same as well 3? What determines the amount of color, the amount present or the concentration in solution? How do wells 3-5 compare to 1 and 2? Record your results.

1.2   To test Le Chatelier's Principle you need to set up an equilibrium and then cause a change to occur and determine the result. The changes that we will measure will be increasing the concentration of reactants and decreasing the concentration of reactants. The former should shift the equilibrium to the right, the latter to the left. In eight wells of your spot plate or well plate add 3 drops each of 0.0025M $Fe(NO_3)_3$ + 3 drops 0.0025M KSCN. Add the following materials to each of the wells, mix each well completely then record your results relative to the control (#6)

Well No.	In each well 0.0025M $Fe(NO_3)_3$ / 0.0025M KSCN (3 drops ea)
6.	Two drops of distilled water (this is the control)
7.	Two drops of 0.0025M $Fe(NO_3)_3$
8.	Two drops of 0.0025M KSCN
9.	Four drops of 0.0025M KSCN
10.	1 drop of 1M KSCN
11.	Two drops of 0.1M Na_2HPO_4
12.	Two drops of 1M H_2SO_4
13.	1 drop of 1M KSCN followed by 3 drops 1M H_2SO_4

In well numbers 7-10 you are adding reactants. Does the red color become more intense -- does the reaction move towards products (the red complex)? In well numbers 11 and 12 you are removing reactants since hydrogen phosphate ion complexes more strongly with iron ion to give a colorless complex and sulfuric acid reacts with thiocyanate ion to give thiocyanic acid which will not complex with iron ion:

$$Fe^{3+} \;+\; HPO_4^{2-} \;\rightleftharpoons\; Fe(HPO_4)^+ \text{ (colorless)}$$

$$SCN^- \;+\; H^+ \;\rightleftharpoons\; HSCN \text{ (will not complex)}$$

The last well (#13) shows the effect of a large excess of SCN- (causing a shift of the equilibrium towards products) followed by a reversal this equilibrium shift towards due to the addition of sulfuric acid.

1.3   A qualitative examination of the solutions used to determine the absorbance of the iron thiocyanate complex will be performed by adding 4 drops of 1M KSCN to each of the wells the followed by the addition to each of five wells varying amounts of $Fe(NO_3)_3$ as described below:

Well No.	In each well 4 drops 1M KSCN
14.	1 drop of 0.0025M $Fe(NO_3)_3$ + 4 drops .1M HNO_3
15.	2 drops of 0.0025M $Fe(NO_3)_3$ + 3 drops .1M HNO_3
16.	3 drops of 0.0025M $Fe(NO_3)_3$ + 2 drops .1M HNO_3
17.	4 drops of 0.0025M $Fe(NO_3)_3$ + 1 drops .1M HNO_3
18.	5 drops of 0.0025M $Fe(NO_3)_3$

Mix each well completely then record your observations. Does the color intensity change from 14-18?

2. Beer's Law Plot for the Absorbance of $Fe(SCN)^{2+}$

2.1   You will perform this portion of the experiment with diluted $Fe(NO_3)_3$ solution and concentrated KSCN (1M). Be sure to use these materials and not the standard $Fe(NO_3)_3$ directly or the more dilute KSCN (0.0025M) solutions.

2.2   Prepare a diluted iron (III) ion solution by using a pipette to transfer 4.00 mL into a 100.00 mL volumetric flask (not an Erlenmeyer flask -- if you do not know what a volumetric flask is check your laboratory handbook or ask the instructor). Dilute this solution to the marked line on the neck of the flask with 0.1 M HNO_3. Invert the flask several times in order to ensure the contents are well mixed.

2.3   Prepare 7 clean and dry test tubes according to the table below. Be sure to use 1M KSCN solution and the diluted $Fe(NO_3)_3$. Once you have added all of the materials stopper and shake each test tube completely to ensure adequate mixing.

Test Tube Number	1.0×10^{-4} M $Fe(NO_3)_3$ mL	1M KSCN mL	0.1M HNO_3 mL
1	1.0	5.0	4.0
2	1.5	5.0	3.5
3	2.0	5.0	3.0
4	2.5	5.0	2.5
5	3.0	5.0	2.0
6	4.0	5.0	1.0
7	5.0	5.0	0.0

2.4   Measure the absorbance (not transmittance) of each solution using the spectrophotometer provided at 450 nm and record this data in the space provided. Be sure to rinse the cuvette (test tube in the spectrophotometer) with 1-2 mL of solution prior to measuring the absorbance.

2.5   Calculate concentration of Fe^{3+} ion in solution for test tubes 1-7 above (for test tube 1 $[Fe^{3+}]$ = 1.0×10^{-4} M * 1.0 mL/10.0 mL). Note that in this case the concentration of Fe3+ in solution is equal to the iron thiocyanate complex in solution (in other words $[Fe^{3+}]initial = [Fe(SCN)^{2+}]eq$, WHY? See Question #1). Plot the absorbance (Y-axis) vs. the concentration of $Fe(SCN)^{2+}$ (X-axis) manually or using a computer program. Determine the slope of the line and record this value in the space provided.

3. Absorbance measurements to determine the Equilibrium Constant

3.1   You will perform this portion of the experiment with 0.0025M $Fe(NO_3)_3$ and KSCN solutions. Do not use the concentrated KSCN (1M) or diluted $Fe(NO_3)_3$.

3.2   Prepare the solutions described in the table below. You may find it convenient to prepare half of the solutions and measure the absorbance and then prepare the rest of the solutions. We will start numbering with test tube 10 in order to ensure that the equilibrium constant data is kept separate from the Beer's Law data.

Test Tube Number	0.0025 M $Fe(NO_3)_3$ mL	0.0025 M KSCN mL	0.1 M HNO_3 mL
10	1.0	1.0	5.0
11	1.0	2.0	4.0
12	1.0	3.0	3.0
13	2.0	1.0	4.0
14	2.0	1.5	3.5
15	2.0	2.0	3.0
16	2.0	2.5	2.5
17	2.0	3.0	2.0
18	3.0	1.0	3.0
19	3.0	2.0	2.0

3.3   Measure the absorbance (not transmittance) of each solution using the spectrophotometer provided at 450 nm and record this data in the space provided. Be sure to rinse the cuvette (test tube in the spectrophotometer) with 1-2 mL of solution prior to measuring the absorbance.

4. Calculation of the Equilibrium Constant Kc

Note: The calculations below can be performed manually or using a spreadsheet computer program such as MS excel. If you use a spreadsheet program indicate using one set of data (one test tube number) how the values were calculated. It will save you time if you calculate the initial concentrations of iron ion for all test tubes and the initial concentrations of thiocyanate ion for test tubes 10-19 prior to the laboratory session.

4.1   Calculate the initial Fe^{3+} and SCN^- concentrations for test tubes 10-19 using the concentration of the reagents used (0.0025M) and their subsequent dilution. For example the Fe^{3+} concentration in test tube 10 is (0.0025M)*(1.0 mL)/(7.0 mL). Record these values in the table provided.

4.2   Calculate the equilibrium concentration of the $Fe(SCN)^{2+}$ complex using the absorbance at 450 nm (A450) for each test tube and the slope determined from the Beer's Law analysis. Use the equation below to solve for the equilibrium concentration of the complex ($[FeSCN^{2+}]eq$).

$$A_{450} = \varepsilon bc = (\varepsilon b)\,[Fe(SCN)^{2+}]eq = (slope)[Fe(SCN)^{2+}]eq$$

4.3   The iron ion (Fe^{3+}) can exist in one of two forms in solution: as the free ion and as the $Fe(SCN)^{2+}$ complex. As a result the sum of the free ion and the complex must equal the total amount of iron present which is the initial iron concentration calculated in 4.1. Therefore it must be true that the iron present as the free ion can be calculated as follows:

$$[Fe^{3+}]eq = [Fe^{3+}] \text{ initial } - [Fe(SCN)^{2+}]eq = \text{Total } Fe^{3+} - \text{complexed } Fe^{3+}$$

Use this information to calculate the free iron ion present at equilibrium ($[Fe^{3+}]eq$) for test tubes 10-19 and enter this information into the table provided.

4.4   The same relationship is true for the thiocyanate ion. Calculate the equilibrium concentration of the thiocyanate ion ($[SCN^-]eq$) for test tubes 10-19 and enter this information into the table provided.

$$[SCN^-]eq = [SCN^-] \text{ initial } - [Fe(SCN)^{2+}]eq = \text{Total } SCN^- - \text{complexed } SCN^-$$

4.5   You now have values for all of the species present in the equilibrium: iron ion, thiocyanate ion, and the complex. Calculate the equilibrium constant for test tubes 10-19 from the equilibrium constant expression.

$$Kc = \frac{[Fe(SCN)^{2+}]eq}{[Fe^{3+}]eq \, [SCN^-]eq}$$

Record the values obtained in the table provided and calculate the average value for all the Kc values obtained. Record this average value in the table provided.

Results: 1. Qualitative examination of equilibria and Le Chatelier's Principle

Well	Contents	drops	Observations
1	$Fe(NO_3)_3$	3	
2	KSCN	3	
3	$Fe(NO_3)_3$ KSCN	1 1	
4	$Fe(NO_3)_3$ KSCN	2 2	
5	$Fe(NO_3)_3$ KSCN	3 3	
6	**$Fe(NO_3)_3$ KSCN water**	**3 3 2**	
7	**$Fe(NO_3)_3$ KSCN $Fe(NO_3)_3$**	**3 3 2**	
8	**$Fe(NO_3)_3$ KSCN KSCN**	**3 3 2**	
9	**$Fe(NO_3)_3$ KSCN KSCN**	**3 3 4**	
10	**$Fe(NO_3)_3$ KSCN 1M KSCN**	**3 3 1**	
11	**$Fe(NO_3)_3$ KSCN Na_2HPO_4**	**3 3 2**	
12	**$Fe(NO_3)_3$ KSCN H_2SO_4**	**3 3 2**	
13	**$Fe(NO_3)_3$ KSCN 1M KSCN then H_2SO_4**	**3 3 1 3**	
14	$Fe(NO_3)_3$ HNO_3 1M KSCN	1 4 4	
15	$Fe(NO_3)_3$ HNO_3 1M KSCN	2 3 4	
16	$Fe(NO_3)_3$ HNO_3 1M KSCN	3 2 4	
17	$Fe(NO_3)_3$ HNO_3 1M KSCN	4 1 4	
18	$Fe(NO_3)_3$ HNO_3 1M KSCN	5 0 4	

2. Beer's Law Plot for the Absorbance of $Fe(SCN)^{2+}$

Test Tube Number	$[Fe^{3+}]_{initial}=[Fe(SCN)^{2+}]_{eq}$	A_{450}
1		
2		
3		
4		
5		
6		
7		

Slope of line ($A450$ vs. $[Fe(SCN)^{2+}]$) $= \varepsilon b \ =$

3. Absorbance measurements to determine the Equilibrium Constant

Slope of Beer's Law Plot from previous page

Test Tube Number	A_{450}	$[Fe(SCN)^{2+}]eq$
10		
11		
12		
13		
14		
15		
16		
17		
18		
19		

4. Calculation of the Equilibrium Constant K_c

Test Tube Number	A_{450}	Initial $[Fe^{3+}]$	Initial $[SCN^-]$	Equil. $[Fe(SCN)^{2+}]$	Equil. $[Fe^{3+}]$	Equil. $[SCN^-]$	K_c
10							
11							
12							
13							
14							
15							
16							
17							
18							
19							

Average value (and SD) for K_c =

DON'T FORGET TO CACLULATE
THE STANDARD DEVIATION OR % ERROR
FOR THE EQUILIBRIUM CONSTANT

Questions:
1. Give a brief definition of Le Chatelier's Principle and how it relates to chemical equilibria.. When you perform the Beer's Law analysis of $Fe(SCN)^{2+}$, it is assumed that the complex concentration is equal to the initial concentration of iron ion added to the solution. Why is this assumption justified? Explain why the data can be analyzed this way including a description of how Le Chatelier's Principle is used in this portion of the experiment.

2. At 720 °C the equilibrium constant Kc for the reaction below is 2.37×10^{-3}. If the concentration of materials is $[N_2] = 0.683M$, $[H_2] = 8.80M$ and $[NH_3] = 3.65M$, predict if the equilibrium will have a net reaction to the right (towards products) or to the left (towards reactants).
$$N_2 + 3H_2 \rightleftharpoons 2NH_3$$

3. What would happen to the A_{450} value for a sample such as test tube #10 if you accidentally added 5 drops of disodium hydrogen phosphate? Explain your answer briefly including any relevant equilibria.

4. You performed these reactions in an acidic solution (diluting with 0.1M HNO_3). This is because iron forms an insoluble hydroxide even when these solutions do not contain excess hydroxide ion (i.e. even in slightly acidic solutions). If the precipitate that forms is $Fe(OH)_{3(s)}$ write the chemical reaction (molecular and full ionic) for the formation of this material starting with $Fe(NO_3)_3$ in aqueous solution. What effect, if any, would you predict using water in place of 0.1M HNO_3 have on the absorbance of this solution at 450 nm for test tube #10?

SCI 124 Principles of Chemistry II
Laboratory 7: Acids and Bases - pH Determination and Buffer Solutions

Purpose:

This laboratory is concerned with the chemistry of acids and bases. The degree of acidity and alkalinity (base content) is normally measured in pH units. This laboratory is focused on the measurement of pH and relating these pH values to acid and base concentrations. The difference between strong and weak acids and bases are examined. Finally the concept of a buffered system will be examined as well as the buffer capacity by observing the pH changes associated with the addition of base to a buffered system. A commercial sample of lemonade is examined to determine if it can be classified as a buffer.

Introduction:

There are many known acids and bases that we come in contact with (though often not directly) in our day-to-day lives. There is strong sulfuric acid in our car batteries and we clean our coffee makers with a 5% solution of acetic acid (white vinegar). The mixing of bleach (chlorine in basic solution) and toilet-bowl cleaner (a strong acid) is known to generate appreciable amounts of chlorine gas. Rolaids are basic in order to react with acid in our stomachs. The measurement of the amount of acid or base in a sample is fundamentally important in the study of chemistry. We could define the acid content of the materials we use in percent or in molarity, but the concentration of protons (H^+ also referred to as hydronium ions H_3O^+) using these types of measurements will change depending on what acid or base is used (note that a 2M sulfuric acid solution is approximately equivalent to a 4M HCl solution). In order to ensure that we have a measurement that reflects the amount of acid dissociated (free protons or hydronium ions) in the solution we measure the pH which is the $-\log_{10}$ of the hydrogen ion concentration. Thus in order to talk about acidity or alkalinity we normally discuss the pH concentration of the solution which is equivalent to the negative logrithmic concentration of hydronium ions.

Water is an acid as well as a base since it can dissociate into hydronium ions (acid) and hydroxide ions (base). The hydronium ions and hydroxide ions formed from the dissociation of pure water are present at relatively low concentrations (1×10^{-7} M) and since the pH of a solution is the $\log_{10}$ of the hydronium ion concentration the pH of pure water is 7.0. Put another way a solution at pH 7.0 is said to be neutral meaning it has no more protons or hydroxide ions than pure water. As the concentration of hydronium ions is increased by a factor of 10 (1×10^{-6} M) the pH would change from 7.0 to 6.0. A similar change in hydroxide concentration from neutral water to ten times as much hydroxide would result in a pH change from 7 to 8 (the hydroxide concentrations would be 1×10^{-7} and 1×10^{-6} M respectively). The normal working limits for aqueous solutions of acids and bases are near zero for strong acids up to +14 for strong bases. Typical acidic solutions will therefore be aqueous solutions with a pH between 0 and 7 and basic solutions will occur with pH values between 7 and 14. It is possible to exceed these limits to give negative pH values (very concentrated strong acid) or pH values greater than 14 (very concentrated strong base) but these are not normally observed. Note that at pH 14 the concentration of hydronium ions will be 1.0×10^{-14} M which is a very low concentration. This laboratory will examine the effects of dilution on the pH value obtained for both acids and bases. The utility of various indicators will also be examined and the pH values that cause a change in color will be determined.

A buffer is a solution that is relatively resistant to changes in pH when diluted or combined with limited amounts of acids or bases. A buffer consists of a weak acid/base conjugate acid-base pair (species that differ by one proton) in which both species are present in appreciable amounts. The functional limits as to what defines a buffer are ratios of the conjugate pair concentrations between 20:1 to 1:20. This means that the species present in the smaller amount must be at least 5% of the total. Buffers can be prepared by adding a strong base to a weak acid provided that the number of moles of base added are less than the moles of weak acid originally present. Likewise adding less than one equivalent of strong acid to a weak base can also form a buffer. In this lab you will prepare three buffer solutions and observe if these solutions resist changes in pH when base is added. Water will also be examined as a non-buffered control (since it has only hydrogen ion and hydroxide ion and no weak acid/base conjugate pair) and lemonade will be examined to see if it behaves like a buffer when treated with base.

Reading Assignments:

You are required to complete the following reading assignments as part of this laboratory session: General Chemistry Atoms First, *2nd Ed.*, McMurry/Fay, Chapter 14 Aqueous Equilibria: Acids and Bases Sections 14.1-14.7 and Chapter 15 Applications of Aqueous Equilibria Section 15.3 Buffer Solutions.

Grading: Grading will be as described in the general laboratory handout.

Procedure:

General:

You will be performing a series of dilutions to generate data both for yourself and the rest of the class. **As a result it is very important to be careful performing these dilutions since contamination by a small amount of highly concentrated acid or base can change the pH values substantially. Use only a new (be sure it _is_ new) disposable pipette to remove samples for testing.** Each dilution will be 10 times less concentrated than the previous solution which for strong acids and bases should theoretically correspond to a 1 pH unit change. You will prepare 25.0 mL of each solution which should be stored in a clearly labeled container.

Procedure

Dilution procedure: Rinse a 25 mL graduated cylinder with distilled water and transfer 10 mL of distilled water to the graduate as measured on the cylinder. Add 2.5 mL of the solution to be diluted to the graduated cylinder and then dilute with water to a total volume of 25.0 mL. This will provide you with a solution with a concentration one tenth as concentrated as the original solution. Subsequent dilution of this sample will dilute by another factor of 10.

You should transfer enough of each solution to a separate clean dry container (test tube, small beaker or well plate) to measure the pH. Be careful not to contaminate your diluted solutions. After obtaining a reasonable pH value (see 1.3 below) be sure to use only the diluted solution to make the next ten-fold dilution. A pH meter and electrode will be available near your work area. Follow the directions of the instructor as to the proper use of the pH meter. The pH values will tend to fluctuate and since we are looking at large changes in pH values record the value if the reading does not change more than 0.1 pH units in 30 seconds. **Do not leave the electrode out of water for extended periods of time. If you are not using the electrode, leave the electrode in a beaker of distilled water.** Be sure to wash the electrode with distilled water before and after each measurement. Set up a 600 mL beaker to serve as a satellite waste receptacle at your work station.

1. Dilution of Acids and Bases. TO SAVE TIME, EACH GROUP (OF 2 OR 3 STUDENTS) SHOULD BE ASSIGNED ONE COMPOUND: ACETIC ACID, HCl, NaOH, or AMMONIA (NH₃) TO DILUTE WITH WATER. DATA SHOULD BE SHARED AND POSTED ON THE BOARD

1.1   Measure the pH of both distilled water and tap water. Record these values in the space provided.

1.2   Four acids and bases will be examined in this part of the experiment. The materials include a strong acid (HCl), a weak acid (HOAc, acetic acid, $HC_2H_3O_2$), a stong base (NaOH), and a weak base (NH₃, ammonia sometimes labelled NH₄OH, ammonium hydroxide). The instructor will assign each group one of these materials for dilution. For the material you are responsible for dilute the solution by a factor of 10 (procedure given above) and place the new solution in a labeled bottle or beaker. Use the diluted solution to further dilute by another factor of ten. This will result in solutions having a range of 10^{-2} M (original solution) to 10^{-7} M (last diluted sample).

1.3   Add enough of the solution you are testing (list above) to a clean dry container (beaker, test tube or well plate) such that the pH electrode (glass part) can be below the level of the solution. This is a 10^{-2} M solution and should be substantially below 7 if an acidic solution and substantially above 7 if a basic solution. Record the pH value for this solution. Note that the pH should be relatively stable but may change over time. If the pH value does not change more than 0.1 pH units over 30 seconds record the value.

1.4   Prepare the first diluted solution and repeat 1.3 for this solution. If the value does not seem reasonable contact the instructor. For strong acids and bases the pH change should be about 1 pH unit. Record the value for the diluted solution and repeat for subsequent dilutions (a total of 5 prepared solutions). The values should increase for the acidic solutions and decrease for the basic solutions. If you feel that the values are not what you would expect contact the instructor. If severe deviations from expected values occur prepare a new diluted solution. Since each dilution is a factor of 10 this should give concentrations from 10^{-2} M (original solution) to 10^{-7} M for the most dilute solutions. Be sure to rinse the electrode with distilled water in between each measurement.

1.5   Add enough of each solution from the original to the most diluted in two rows of a well plate (long edge). To the top row add one or two drops of universal indicator. To the second row add on or two drops of bromothymol blue (if the original solution was an acid) or phenolphthalein (if the original solution was a base). Record your observations as to how the dyes changed as the solutions become more dilute.

1.6   Post your results (pH and color change of both indicators) on the board for other students to use. Record the other group's results for the three solutions that your group did not test. This means that you should have dilution data for the four original solutions that include how the pH and indicator colors change as the solution is diluted from 10^{-2} to 10^{-7} M.

1.7   Dispose of the solutions in the wells using a 600 mL beaker at your work station and rinse the 24 well tray with distilled water. Use a paper towel to dry each well (a small amount of water remaining is acceptable). **Make sure that you have recorded all of the pH values and color changes in the table provided before you discard your solutions.**

2. Buffer Preparation and Buffer Capacity. TO SAVE TIME EACH GROUP SHOULD BE ASSIGNED A BUFFER A, B, OR C AND SHARE DATA. EVERY GROUP NEEDS TO DO WATER AND LEMONADE TITRATION.

General: You will prepare one of three buffer solutions: Buffer A (acetic acid and sodium acetate in a 1:1 molar ratio), Buffer B (acetic acid and sodium acetate in a 1:4 molar ratio), and Buffer C (ammonia and ammonium chloride in a 1:1 molar ratio). Your instructor will indicate which buffer solution your group will be responsible for preparing and measuring. You will record buffer capacity data for your buffer, water and lemonade. Be sure to record your data on the board such that other groups can collect and interpret this data.

2.1 In a clean, relatively dry small beaker or large test tube combine solution 1 and solution 2 as described in the table below and label your container as Buffer A, B or C.

Buffer	Solution 1	Solution 2
A (1:1 Acetic Acid:Sodium Acetate)	7.5 mL of 10^{-2} M acetic acid	7.5 mL of 10^{-2} M sodium acetate
B (1:4 Acetic Acid:Sodium Acetate)	3.0 mL of 10^{-2} M acetic acid	12.0 mL of 10^{-2} M sodium acetate
C (1:1 Ammonia:Ammonium Chloride)	7.5 mL of 10^{-2} M ammonia	7.5 mL of 10^{-2} M ammonium chloride

2.2   Record the pH of your solution in the data sheet provided.

2.3   Set up a buret containing 10^{-3} M NaOH solution (remember to rinse the buret with solution prior to filling the buret). Add 30 mL of distilled water and a magnetic stir bar to a 250 mL beaker. Place the beaker on a stir plate under the buret tip and start the stirrer.

2.4   Place the pH electrode into the solution such that the electrode tip is completely submerged in the water but such that it does not interfere with the stir bar. Record the pH of the solution.

2.5   Slowly add the NaOH solution from the buret until the pH changes by at least 1.0 pH unit (for example from pH 6.5 to pH 7.5). If the pH changes less than 1.0 unit then add a maximum of 50.0 mL of sodium hydroxide (one full buret). Record the final pH and the volume of NaOH solution added. Do not add so much NaOH solution that the level falls below the markings on the buret.

2.6   Dilute the buffer solution prepared in 2.1 by adding 15.0 mL of water to your 15.0 mL of buffer solution. Record the pH of the diluted buffer solution in the data sheet provided

2.7   Repeat 2.4-2.5 for this diluted buffer solution as you did for water. Record the volume of NaOH added to give a 1.0 pH unit change or the volume added from the buret that can be measured (50.0 mL or less). Record the pH of the solution after the addition of the sodium hydroxide.

2.8   Repeat 2.7 for lemonade (15 ml lemonade combined with 15 mL of water). NOTE YOU MAY WANT TO DILUTE THELEMONADE MORE BEFORE USE B/C 1:1 REQUIRES > 50 ml OF NAOH FOR 1 PH UNIT CHANGE. TRY 10 ML LEMONADE IN 2O ML OF WATERBUT MAKE SURE TO NOTE 1:3 NOT 1:1 DILUTION LIKE THE OTHERS.

2.9   Determine the change in pH per unit volume by subtracting the final pH from the initial pH and dividing this quantity by the volume of NaOH that was added. This should give a value that can be used to measure the ability of these solutions and water to resist changes in pH when strong base has been added.

CALCULATIONS FOR DILUTION OF ACIDS AND BASES HERE:

Results: **1. Dilution of Acids and Bases**

pH of distilled water is _____   pH of tap water is _____

Conc.	1. 10^{-2} M	2. 10^{-3} M	3. 10^{-4} M	4. 10^{-5} M	5. 10^{-6} M	6. 10^{-7} M
A. HCl **pH**						
Color (Universal)						
Color (BT Blue)						
B. Acetic Acid **pH**						
Color (Universal)						
Color (BT Blue)						
C. NaOH **pH**						
Color (Universal)						
Color (Phth.)						
D. NH_3 **pH**						
Color (Universal)						
Color (Phth.)						

2. Buffers and Buffer Capacity

	Undiluted pH [2.2]	Initial (diluted) pH [2.4, 2.6]	Final pH [2.5, 2.7]	Volume of NaOH added	ΔpH/ Δ Volume [2.9]
Buffer A [A⁻]/[HA] = 1:1					
Buffer B [A⁻]/[HA] = 4:1					
Buffer C [BH⁺]/[B] = 1:1					
Distilled Water					
Lemonade					

Questions:

1. Calculate the number of grams of NaOH per 100 mL to give a pH 8.0 solution if pure water at pH 7.0 is used in the experiment. Repeat this calculation for a pH 6.0 solution using HCl. Comment on how this would or would not explain the fact that distilled and tap water are not exactly at a pH value of 7.000.

2. Compare the pH values in the dilution experiment for each of the acids and bases used to the pH determined for distilled water. Do the pH values for the more dilute solutions approach the pH of the distilled water? What would you expect to happen if you continued to dilute these samples even further? Explain your answer.

3. Compare the pH values in the dilution experiment for the strong acid (HCl) and the weak acid (acetic acid). Do the pH values increase at the same rate across the wells?

4. What would you expect to happen if you combined 10.0 mL of 10^{-2} M acetic acid and 5.0 mL of 10^{-2} M sodium hydroxide? What would be the concentrations of acetic acid and sodium acetate in this solution? Would you expect this solution to be a buffer? Explain your prediction.

5. Compare the $\Delta pH / \Delta$ Volume (the pH change divided by the volume of NaOH added) values for each of the solutions and water in part 2. Is there any difference between Buffer A and Buffer B when you compare these values? Would you classify lemonade as a buffer from this data? Give a brief explanation of your classification of lemonade.

SCI 124 Principles of Chemistry II
<u>Laboratory 8: pH Titration Curve and MW of an Unknown Acid</u>

<u>Purpose:</u>

This laboratory will introduce the student to how pH measurements change during a titration. The student will generate pH titration curves and compare the endpoint determined by a titration curve to that obtained using an indicator. The shape of the pH titration curve for a strong acid (HCl) will be compared to the pH titration curve for a weak acid (acetic acid). The technique of using a derivative curve to locate the endpoint will also be demonstrated. Students will also identify if an acid is monoprotic, diprotic or triprotic by analyzing the derivative curve of an unknown acid. Finally, the molecular weight of this unknown acid will be determined by titration.

<u>Introduction:</u>

The use of titration is one of the most common techniques used to measure solution concentration in the laboratory. Titration is a useful method provided that the materials used react completely and quickly. This is almost always the case for acid-base reactions where hydroxide ions and hydronium ions (hydrated protons) combine to form water in a neutralization reaction. The quantitative application of these acid-base reactions involves the ability to accurately measure the volume of a solution required to neutralize the acid or base that is being titrated. A buret is used to accurately measure the amount of acid or base used to neutralize the analyte. It is also necessary to determine at what point all of the acid has reacted. The pH must change during the titration and if one is titrating a strong acid with a strong base the pH must change from a low value to a high value. This change is not linear, however, and the shape of the titration curve (a graph of pH vs. volume of base) can be used to accurately determine the endpoint. The use of this method will be compared to the use of an indicator which changes color as a function of pH to determine the neutralization point or endpoint of the titration.

If you start with a highly pure material it is straightforward to prepare a solution of known concentration. If the mass of a pure substance is determined accurately and is subsequently diluted to a known final volume, the concentration in moles/liter (molarity, M) can be calculated by converting the mass of substance to moles and using the final volume of the solution prepared. Sodium hydroxide is **not** one of the materials for which this method can be used. The reason for this is that the amount of water contained in sodium hydroxide pellets is substantial (5-10%) and will vary depending on how the solid is stored, how long it has been opened, and a number of other variables that are difficult to control. As a result, the concentration of sodium hydroxide solutions <u>are not determined by calculating the moles present in a given mass of sodium hydroxide pellets.</u> Instead the approximate concentration is prepared and the solution is then measured to determine its concentration. This procedure is called the **standardization** of a solution. In this case we will use a solid which is highly pure, has a known structure, and contains only one acidic proton to react with sodium hydroxide. The material used to standardize the NaOH solution is potassium hydrogen phthalate (KHP). A titration of a known (measured) mass of KHP with the unknown concentration of sodium hydroxide will give a volume of sodium hydroxide solution required to neutralize the KHP. Since we know the number of moles of KHP, the ratio of moles of NaOH to KHP (1:1), and the volume of sodium hydroxide solution required, we can calculate the concentration of the sodium hydroxide solution in moles/liter (M). Repetition of this procedure by other groups in the laboratory will give an average value for the sodium hydroxide solution molarity (concentration). When this method is used correctly the volumes of reagents transferred by buret are accurate to within 0.02 mL and the concentration of the solutions are accurate to at least three significant figures (ex. 0.264M).

The graphical analysis of pH as a function of volume of base added in the titration of an acid can be used to determine the endpoint. The pH change is relatively small until the volume of base is nearly equal to the amount required to neutralize the acid. Thus the pH changes very rapidly at the end point(the vertical slope of the graph is nearly zero) and at volumes of base past the endpoint the rate of change is also lower since once the acid has been exhausted, the pH change will be that for adding additional base to a basic solution(the horizontal slope approaches zero). The maximum rate of change will be the endpoint of the titration. This can be determined graphically or by generating a "first derivative" curve which can be obtained by graphing the change in pH vs. the volume of base added. The maximum point in this graph is a more accurate means of determining the volume of base added at the endpoint. **The number of peaks on the derivative curve is directly linked to the type of acid (monoprotic, diprotic or triprotic) being used**. It will also be noted that the pH curve for a weak acid has a slightly different shape from that for a strong acid. These titration methods will be used to determine the molecular weight of an unknown acid that will be provided to you.

Reading Assignments:

You are required to complete the following reading assignments as part of this laboratory session: General Chemistry Atoms First, *2nd Ed.*, McMurry/Fay, Chapter 15 Applications of Aqueous Equilibria Sections 15.7 Weak Acid-Strong Base Titrations and 15.9 Polyprotic Acid-Strong Base Titration

Procedure:

1. Preparation and standardization of Sodium Hydroxide Solution.

Warning: Sodium hydroxide can cause chemical burns !! Clean up any spilled sodium hydroxide pellets and use care when working with the sodium hydroxide solution. A slippery or slimy feel on your skin is a good indication that you have come into contact with sodium hydroxide. Alert the instructor and wash the affected area completely with water.

1.1   One pair of students will prepare enough sodium hydroxide solution for the entire laboratory section (2.00L). The concentration should be 4-5 grams of sodium hydroxide per liter of solution (0.1 M).

1.2   While the sodium hydroxide solution is being prepared the other students should set up a titration apparatus including a buret, stir plate, a pH electrode/iBook computer, Lab Pro system and a 250 mL beaker. Set up the iBook as far from the buret and beaker as possible to avoid damaging the computer.

1.3   Each pair of students will perform a standardization of the sodium hydroxide solution using KHP (potassium hydrogen phthalate). The data for the standardizations will be written on the blackboard. You should use significant figures correctly when you provide your data to the class. The average concentration of the sodium hydroxide will be determined by the class. Be sure to record the average value for the sodium hydroxide solution during the laboratory session.

1.4   Weigh out approximately 0.4 grams of potassium hydrogen phthalate (KHP) on an analytical balance and record the mass to within 0.1 mg (0.0001 grams). Transfer the KHP to a 250 mL beaker containing a magnetic stir bar (rinse flask and stir bar with distilled water before transferring the KHP).

1.5   Add approximately 50 mL of distilled water to the flask and add 2-3 drops of phenolphthalein solution to the beaker. Wash any solid KHP or phenolphthalein solution on the sides of the flask with a small quantity of distilled water from a wash bottle.

1.6   Transfer some of the sodium hydroxide solution which was prepared to a clean and dry 50 mL beaker and wash the buret with 3-6 mL of solution three times (be careful handling the sodium hydroxide solution!). Place the buret into the buret clamp and fill the buret to exactly the highest volumetric mark. If not at zero record the volume as Vo using appropriate units. You should be able to read the buret volume to within 0.02 mL. Be sure to record the volume where the meniscus (lowest point in the liquid column) meets the markings on the buret.

1.7   Start the magnetic stir bar in the flask with the buret tip inside the beaker such that the NaOH solution will be added directly to the KHP solution. Add the NaOH solution by opening the stopcock. Once you see a definite pink color in the solution add the NaOH solution in smaller volume increments and finally dropwise. You want to obtain the volume of NaOH necessary to change the color of the KHP solution to a light pink that does not fade. You can use the wash bottle of distilled water to wash down the sides of the beaker and the tip of the buret as needed. If performed correctly the endpoint can be reached with only one drop (or fraction of a drop) of NaOH solution. You can add a partial drop by opening the stopcock slightly to form a ball of liquid at the end of the tip (but not enough for a full drop to fall) and washing the NaOH solution from the tip using distilled water. Record the volume required to produce the expected pink color (endpoint of the titration).

1.8   Calculate the number of moles of KHP used, the volume of sodium hydroxide required (the final volume minus the initial volume if not zero) to reach the endpoint and from these data the sodium hydroxide solution concentration in molarity (M). Again use significant figures correctly in this calculation.

1.9   Write your concentration of sodium hydroxide on the blackboard and record the NaOH solution concentration obtained by the other groups. Determine the average value for the sodium hydroxide concentration and use this value in all subsequent calculations.

2. Determination of the Molecular Weight of an Unknown Acid.

2.1   Obtain a sample of unknown from the instructor's bench or lab cart and record the number of the unknown on the data sheet provided.

2.2   If you pick unknown 1 or 2, weigh a 0.2-0.30 gram sample of the unknown and set up the titration in the same way that you did for the KHP standardization (1.4). Do a continuous titration curve as you did in 3.2

        If you pick unknown 3 or 4, weigh a 0.15-0.20 gram of unknown and set up the titration in the same way that you did for the KHP standardization (1.4). Add about 14 drops of 6M HCl to lower the pH of the solution to about 1.87 before titrating with NaOH. You will record the titration curve as a continuous time graph using the iBook/LabPro system. You will need to record the total volume of NaOH used in the experiment, the total time of the data acquisition and to open/close the stopcock of the buret at the same time you start/stop the data acquisition on the computer.

2.3   In order to automatically record the titration curves select Experiment from the pull-down menu and select Data Collection. Change the Length window to read 4 minutes and the Sampling Speed to 60 samples per minutes (0.016666 min per sample). Select DONE in the bottom right corner of the dialogue box.

2.4   Fill the buret to the zero mark and record the initial volume of NaOH (or 0.00 mL if adjusted exactly to zero). Start the stir bar and position the electrode as you did in part 2 for HCl.

2.5   Simultaneously press the Start button above the graph window and open the stopcock to allow the sodium hydroxide solution into the beaker at a constant rate (but not completely open). Observe the pH values on the graph as they appear and stop the acquisition by **both pressing the Stop button above the graph window and closing the stopcock** when the pH exceeds 12 or the volume of sodium hydroxide in the buret is near the bottom of its graduated region (where there are volume markings). Be sure that you record the volume of sodium hydroxide in the buret after the stopcock has been closed as well as the time at which the addition was stopped (last time value in data table)

2.6   To obtain the first or second derivative of a curve using the iBook/DataLogger Pro select Data/New Calculated Column (you can enter a name if you wish and it will identify the graph on the y axis). Under Functions select Calculus and Derivative. This should give a window that displays Equation: derivative (|). Software collect the data (pH vs. time for example) in the standard manner. To select the data for the derivative enter pH as the first (y) variable, enter a comma, and enter Time as the second (x) variable. This should give a display in the window of Equation: Derivative ("pH", "Time"). Select Done from the lower corner of the window. Insert/Graph should display the derivative curve and you can use Analyze/Examine to place the cursor at the max before printing. You can select print from the draw-down menu under File or Apple P to print. The same procedure can be followed in the second derivative curve is desired. For this curve the zero cross-over point would represent the endpoint (going from positive to negative slope in the first derivative curve).

2.7   Repeat this procedure two more times using slightly more or less of the unknown depending on how much base was required (you want to keep the endpoint greater than 15 mL but less than 40 mL).

2.8   Using the concentration of sodium hydroxide, the volume of sodium hydroxide from the endpoints determined (this would be the point of greatest change of pH or the max of the first derivative curve or the zero point of a second derivative curve), and the mass of the unknown acid added calculate the molecular weight of the unknown acid and determine the average molecular weight. Examine the pH titration curve for evidence of more than one endpoint. Note that your endpoint data will give you the time at the endpoint. To determine the volume of NaOH at that time multiply the total volume of NaOH added by the ratio of the time at the endpoint to the total time ($V_{endpoint} = V_{total} \times$ (time at endpoint/total time))

Figure 1. Determination of Endpoint from Titration Curve

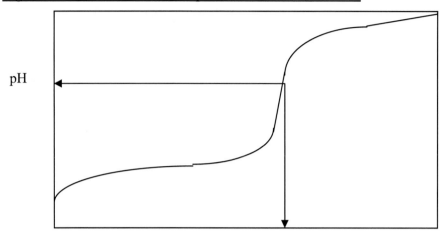

Volume of Titrant (NaOH)

Chemical Reactions:

1. KHP Standardization

$$HKC_8H_4O_4 + NaOH \longrightarrow NaKC_8H_4O_4 + H_2O$$

(KHP) Note: KHP has a MW of 204.2 g/mole and only one acidic proton

2. Acetic acid reaction

$$HC_2H_3O_2 + NaOH \longrightarrow NaC_2H_3O_2 + H_2O$$

Results:

1. Preparation and standardization of Sodium Hydroxide Solution.

Mass of KHP	Volume of NaOH at endpoint
Moles of KHP (FW = 204.2 g/mole)	Molarity of NaOH
Molarity of NaOH from other groups	
Average NaOH molarity	

2.   **Determination of the Molecular Weight of an Unknown Acid.**

Unknown ID number:

	Trial 1	Trial 2	Trial 3
Mass of acid			
Total volume NaOH			
Total Time			
Time at endpoint			
Volume at endpoint			
moles of NaOH			
moles of acid (More than one proton?)			
MW of acid			
Average MW of trials			

Questions:

1. Compare the three methods of determining the endpoint of the titration used (indicator, graphical, and derivative). Do all three methods give the same result (the same volume of NaOH at the endpoint)? What are the differences and which do you feel is the most accurate.

2. Compare the pH vs. volume curve for HCl and acetic acid shown in the textbook. Discuss any Differences you see in the shape of the curves. Compare these two curves to one of the curves prepared for the unknown. Is the unknown acid more similar to HCl or acetic acid?

3. Examine the pH vs. volume curve for the titration of your unknown. Is there any evidence of more than one acidic proton on you unknown? If not then what would the graph of an unknown that has more than one equivalence point look like? Calculate the molecular weight (equivalent weight) of your unknown.

4. Suppose your unknown was a white solid which showed two equivalence points in the titration curve. If 0.2114 grams of unknown required 38.44 mL of 0.106 M NaOH to be completely neutralized, what is the molecular weight of this compound. Be sure to include the fact that there were two equivalence points in your answer.

Principles of Chemistry II
Laboratory 9: Inorganic Qualitative Analysis

Purpose:

Inorganic qualitative analysis is a classical method of separating cations and anions from complex mixtures by utilizing the similarities in solubilities for groups of ions. This laboratory will investigate the separation of cations using selective conditions to isolate cations in a mixture based upon their solubility characteristics. Representative cations from Analytical Group 1 (insoluble chlorides), Group 2 (insoluble sulfides in acidic solution), Group 3 (insoluble sulfides in basic solution) and Group 4 (insoluble carbonates or phosphates) will be separated and their identity confirmed by an additional test. The identity of an unknown which contains 1-3 of the possible four cations will be determined by using this qualitative analysis scheme. The concepts of solubility and complex ion formation will be presented in this laboratory session.

Introduction:

There are a variety of ways to separate components of a mixture. You have some experience separating organic materials using chromatography. Since most inorganic materials are ionic and water soluble, the methods required for the separation of ionic materials are usually performed in aqueous solution. It is possible to separate ions by a chromatographic method called ion exchange chromatography. This laboratory will examine how to separate ions in a mixture due to solubility differences. This is often called a "wet" method since it requires chemical transformations and isolation as opposed to those performed by instruments. Inorganic qualitative analysis is a classical wet method which can be applied to cations or anions in solution. The separation and identification of these materials arises from the selective solubility of groups of ions and confirmation of identities from known chemical reactions.

Our examination of inorganic qualitative analysis will include four groups of cations which we will call Analytical Groups 1, 2, 3, and 4 (McMurry/Fay Section 15.14 and 15.15). The separation of multiple ions in each group is a much more difficult experiment so we will examine one ion which will be representative of each group. Each group will form a precipitate under different conditions as described in the table below. If we perform these precipitation reactions in a specific order or sequence then we can remove one group from a mixture containing more than one group. For example, Group 1 cations can be precipitated by treatment with HCl since they all form insoluble chlorides but Group 2, 3, and 4 cations do not react with HCl. As a result the precipitate formed from treating a mixture of Group 1, 2, 3 and 4 ions with HCl will remove **only** Group 1 cations as insoluble chlorides leaving the other cations in solution available for further testing.

Analytical Group	Cations in Group	Conditions of Isolation	Representative Cation
Group 1	Ag^+, Hg_2^{2+}, Pb^{2+}	Treat with HCl Precip. Chloride	Ag^+ (AgCl)
Group 2	As^{3+}, Bi^{3+}, Cd^{2+}, Cu^{2+} Hg^{2+}, Pb^{2+}, Sb^{3+}, Sn^{4+}	Treat with H_2S/H^+ Precip. Sulfide	Cu^{2+} (CuS)
Group 3	Co^{2+}, Fe^{2+}, Mn^{2+}, Ni^{2+} Zn^{2+}, Al^{3+}, Cr^{3+}	Treat with H_2S/OH^- Precip. Sulfide	Zn^{2+} (ZnS)
Group 4	Ba^{2+}, Ca^{2+}, Mg^{2+}, Sr^{2+}	Treat with CO_3^{2-} Precip. Carbonate	Ca^{2+} (CaCO$_3$)

For your laboratory session you will examine a known that contains Ag^+, Cu^{2+}, Zn^{2+} and Ca^{2+} ions (representing Group 1, 2, 3, and 4 respectively) and an unknown that contains 1-3 of these ions. You will isolate each cation and perform an additional test in order to confirm the identity of that ion from a known reaction for each of the cations.

You might notice in the table above that both Group 2 and Group 3 cations (in this laboratory Cu^{2+} and Zn^{2+}) both form insoluble sulfides. How is it possible to separate these two ions if they both precipitate in the presence of sulfide ion (S^{2-})? It turns out that although both copper and zinc ions form an insoluble sulfide the copper compound is far less soluble than zinc compound. Furthermore if you look at the solubility equilibria you will find that the solubility of these types of compounds are defined by a solubility equilibrium constant called the Ksp (McMurry/Fay Chapter 15 Sections 15.10 and Table C.4 in Appendix). If you look up the Ksp values for CuS and ZnS you will observe that the CuS value is much smaller than the ZnS value. As a result if you control the concentration of sulfide ion (S^{2-}) you can remove copper from zinc due to the solubility differences in the sulfides. Controlling the sulfide ion concentration is accomplished by adjusting the pH at which the precipitation occurs. In highly acidic solutions (pH 0.3) the sulfide ion concentration is much lower than in basic solutions (pH 9) due to the acid ionization equilibrium of H_2S. Thus at low pH only CuS will precipitate out and at high pH both CuS and ZnS would be removed. This allows you to selectively remove copper ion from a mixture which also contains zinc ion.

For both Group 2 and 3 cations you will be separating these as sulfides by forming H_2S in solution. The H_2S is formed from the reaction of an organic compound called thioacetamide which when heated decomposes to form H_2S in solution. You should be aware that thioacetamide is a hazardous chemical and should be treated with great care. The H_2S that you form in solution will be a saturated solution ($[H_2S] = 0.1M$) but pure H_2S is a foul smelling gas which is considerably toxic. Be careful with both thioacetamide and hydrogen sulfide in this laboratory session.

Reading Assignments: You are required to complete the following reading assignment as part of this laboratory session: Chapter 15 Applications of Aqueous Equilibria Section 15.11 – 15.15 Precipitation of Ionic Compounds, Section 15.14 Separation of Ions by Selective Precipitation, and Section 15.15 Qualitative Analysis.

Grading: Grading will be performed as described in the general laboratory handout, but the identity of the unknown will be worth 10 points).

Procedure:

General: You should keep a 600 mL beaker at your work station to use as a satellite waste receptacle but do not discard solutions until you are certain that you don't need them anymore. **Always perform the experiments on your unknown and known mixture simultaneously in order to ensure that you have used the appropriate conditions to obtain a positive test for the known.** Hydrogen sulfide (H_2S) which you will form *in situ* (which means you will prepare it in solution from a chemical reaction) is both foul smelling and toxic. Although only small quantities of H_2S are liberated in the air, work with the thioacetamide solutions in one of the fume hoods if you feel that the liberation of H_2S is excessive. Thioacetamide is a hepatotoxin and carcinogen so be careful using this solution and wash your hands thoroughly after handing thioacetamide. You can estimate volumes by assuming that 20 drops is equal to 1.0 mL. Each test will have an identification section (ID) and a confirmation section (Confirm.).

1. Test for Silver (I) Ion (Group 1): [ID 1.1-1.4]; [Confirm. 1.5-1.6]

1.1   Obtain a sample of the known mixture of all four cations and a sample of your unknown. Record the number of your unknown on your laboratory sheet. Use a heating block set to give a constant temperature just below 100°C (around 85-90°C). **Use only test tubes that will fit in the centrifuge and heating block.**

1.2   Add 30 drops of your known and unknown to two separate marked test tubes. Add 2 drops of 6 M HCl (CAUTION! Strong acid) to each of the test tubes and agitate each of the test tubes to mix the contents.

1.3   If a precipitate forms (and it should for the known) centrifuge the test tube and add 1 drop of 6 M HCl to test if the precipitation was complete. If more solid forms from the additional HCl centrifuge again and repeat until treatment of the solution with HCl does not give any additional precipitate. If NO precipitate formed in your unknown sample save this solution further testing in section 2.1.

1.4   Centrifuge test tube if necessary and remove the solution above the precipitate from above using a disposable pipette or carefully pour off the solution (decant the solution) to separate the liquid layer. Save this solution for further testing in section 2.1.

1.5   Ammonia will dissolve silver chloride by complexing the silver ion. In order to confirm the presence of silver ion in the precipitate take the solid from 1.4 and treat with 6 drops of 6 M ammonia (NH_3). Agitate the test tube to ensure that the contents of the test tube are thoroughly mixed. If some of the solid dissolved but not all add another 3 drops of ammonia solution.

1.6   If your precipitate was only AgCl it should dissolve in ammonia solution. If some of your sample dissolved centrifuge the sample and remove only the solution. Treat this solution with 6M HNO_3 (CAUTION! Strong Acid/Oxidizer) until the precipitate reappears. If the solid does not appear for the known contact the instructor. The dissolution of the precipitate from treatment with ammonia and the reappearance of the precipitate from treatment with nitric acid **confirms the presence of silver ion.** If your unknown exhibited this behavior then it contains silver ion as one of its constituent ions. If your known and unknown behaved the same in these tests then record a positive test for silver ion in your unknown.

2. Test for Copper (II) Ion (Group 2): [ID 2.1-2.5]; [Confirm. 2.6-2.9]

2.1   You should have two solutions for further testing at this point: one for the known (from 1.4) the other for the unknown (from 1.3 or 1.4). Take each of these solutions and add 6 M NH_3 drop wise (agitating the test tube after each addition) until the solution is basic by litmus paper or pH paper.

2.2   In order to separate only Group 2 cations (not Group III) you need to carefully adjust the acidity of the solution prior to the next step. Estimate the volume in each of the test tubes and increase this value by 0.6 mL in order to account for the dilution of thioacetamide in 2.3. Using this estimated volume add 1 drop of 6M HCl per milliliter of solution to each of the test tubes. This should provide a solution with a pH near 0.5 ($[H^+] = 0.3$ M approximately). It is not important that the pH be exactly 0.5 but it must be appreciably acidic. Test the solution with wide range pH paper and ensure that the pH is less than 2.0. If the pH is greater than 2.0 add 6 M HCl drop wise until it is below 2.0 as measured using wide-range pH paper.

2.3   Add 12 drops (0.6 mL) of 1 M thioacetamide to each of the solutions **(Be careful handling thioacetamide!)** and place the test tubes in the hot water bath for 8-10 minutes.

2.4   You should observe a black precipitate for your known sample (CuS) and you may or may not observe a similar precipitate for your unknown. If a precipitate is observed centrifuge the test tube and separate the solution from the solid. If no precipitate was observed for your unknown save this solution for further testing in section 3.1.

2.5   Take the solution centrifuged in 2.4 and treat with a additional 3 drops of thioacetamide to ensure complete precipitation of all of the copper ion present. Place this test tube in the hot water bath for five minutes. If additional precipitate appears centrifuge the solution and save the liquid for further testing in 3.1.

2.6   Take the precipitate obtained in 2.4 and combine the precipitate formed in 2.5 (if that quantity of solid is appreciable). Add 1-2 mL of distilled water and mix the contents thoroughly. Centrifuge the test tube and discard the solution (saving the solid).

2.7   Treat the solid from 2.6 with 1.0 mL (20 drops) of 6 M HNO_3 (CAUTION! Strong Acid/Oxidizer). Place this test tube in the hot water bath for 2-3 minutes (be sure that the test tube is not pointing at anyone!) with occasional stirring of the test tube. Centrifuge this test tube and retain the solution for the next step.

2.8   Add 6 M NH_3 to the solution obtained in 2.7 until it is basic by litmus or pH paper. Add 6 M acetic acid (HOAc) until the solution is acidic by litmus or pH paper.

2.9   Take the solution from 2.8 and add 0.5 mL (10 drops) of 0.1 M potassium ferrocyanide

($K_4Fe(CN)_6$) solution. Agitate the test tube to thoroughly mix the contents. The presence of a red-brown precipitate **confirms the identity of copper (II) ion** in the sample. If you obtained a black precipitate from the thioacetamide but no precipitate in 2.9 heat the test tube for 3 minutes to observe if a precipitate forms.

3. Test for Zinc Ion (Group 3): [ID 3.1-3.4]; [Confirm. 3.5-3.6]

3.1   You should have two solutions for further testing at this point: one for the known (from 2.5) the other for the unknown (from 2.4 or 2.5). Take each of these solutions and add 0.5 mL (10 drops)

of 6 M HCl then add 6 M NH_3 drop wise (agitating the test tube after each addition) until the solution is basic by litmus paper or pH paper.

3.2   Add an additional 5 drops of 6 M NH_3 to this solution and check the pH using pH paper. The pH of this solution should be appreciably basic with a pH greater or equal to 9. If the pH is less than

9, add additional 6 M NH_3 until it is at least pH9 by pH paper.

3.3   Add 12 drops (0.6 mL) of 1 M thioacetamide to each of the solutions **(Be careful handling thioacetamide!)** and place the test tubes in the hot water bath for 8-10 minutes.

3.4   You should observe a white or near white precipitate for your known sample (ZnS) and you may or may not observe a similar precipitate for your unknown. If a precipitate is observed centrifuge the test tube and separate the solution from the solid. If no precipitate was observed for your unknown save this solution for further testing in section 4.1.

3.5   Take the solution centrifuged in 3.4 and treat with a additional 3 drops of thioacetamide to ensure complete precipitation of all of the zinc ion present. Place this test tube in the hot water bath for five minutes. If additional precipitate appears centrifuge the solution and save the solution for further testing in 4.1.

3.6   Combine the precipitate obtained in 3.4 with the precipitate formed in 3.5 (if that quantity of solid

is appreciable). Add 1-2 mL of distilled water, 2 drops of ammonia (NH_3) and mix the contents thoroughly. Centrifuge the test tube and discard the solution (saving the solid).

3.7   Treat the solid from 3.6 with 1.0 mL (20 drops) of 6 M HNO_3 (CAUTION! Strong Acid/Oxidizer). Place this test tube in the hot water bath for 2-3 minutes (be sure that the test tube is not pointing at anyone!) with occasional stirring of the test tube. Centrifuge this test tube and retain the solution for the next step.

3.8   Add 6 M NH_3 to the solution obtained in 3.7 until it is basic by litmus or pH paper. Add 6 M acetic acid until the solution is acidic by litmus or pH paper.

3.9   Take the solution from 3.8 and add 0.5 mL (10 drops) of 0.1 M potassium ferrocyanide

$(K_4Fe(CN)_6)$ solution. Agitate the test tube to thoroughly mix the contents. The presence of a white precipitate **confirms the identity of zinc (II) ion** in the sample. If you obtained a precipitate from the thioacetamide treatment but no precipitate in 3.9 heat the test tube for 3 minutes to observe if a precipitate forms.

4. Test for Calcium Ion (Group 4): [ID 4.1-4.2]; [Confirm. 4.3-4.5]

4.1   You should have two solutions for further testing at this point: one for the known (from 3.5) the other for the unknown (from 3.4 or 3.5). Take each of these solutions and add 0.5 mL (10 drops)

of 3 M ammonium carbonate $((NH_4)_2CO_3)$. A white precipitate indicates the presence of calcium ion. This should occur for the known sample and may or may not occur for the unknown.

4.2   If a precipitate forms in 4.1 centrifuge the solution and separate the solid. Add 1.0 mL (10 drops) of water to the solid and mix the contents thoroughly. Centrifuge the sample again and separate the solid.

4.3   Dissolve the solid in 4.2 with 5 drops of 6 M acetic acid. If the solid doesn't completely dissolve add a few more drops of acetic acid.

4.4   Make this solution basic by litmus paper or pH paper by adding 6 M ammonia (NH_3). Add 0.5

mL (10 drops) of 0.1 M potassium ferrocyanide $(K_4Fe(CN)_6)$ solution. Agitate the test tube to thoroughly mix the contents. The presence of a white precipitate **confirms the identity of calcium (II) ion** in the sample. If no precipitate forms heat the solution 3 minutes in the water bath.

4.5   You may be instructed to substitute 1M potassium oxalate (CAUTION! Poison, $K_2C_2O_4$) for the potassium ferrocyanide solution. If this is the case a white precipitate also **confirms the identity of calcium (II) ion** in the sample.

5. Confirmation and testing of the unknown identity

5.1   It is important that you **record the unknown number** of the unknown that you used. It should be possible from the tests above to identify all of the ions present in your unknown. Remember to compare your unknown results directly to the known solution. If there are identification or confirmatory tests that appear ambiguous contact the instructor and perform the same test on a sample that has only the cation of interest. **Be sure to record your identification of the ions present in the unknown sample.**

Results:
WHAT IS YOUR UNKNOWN NUMBER? _____

Group 1: Test for Silver (I) ion (Ag^+)

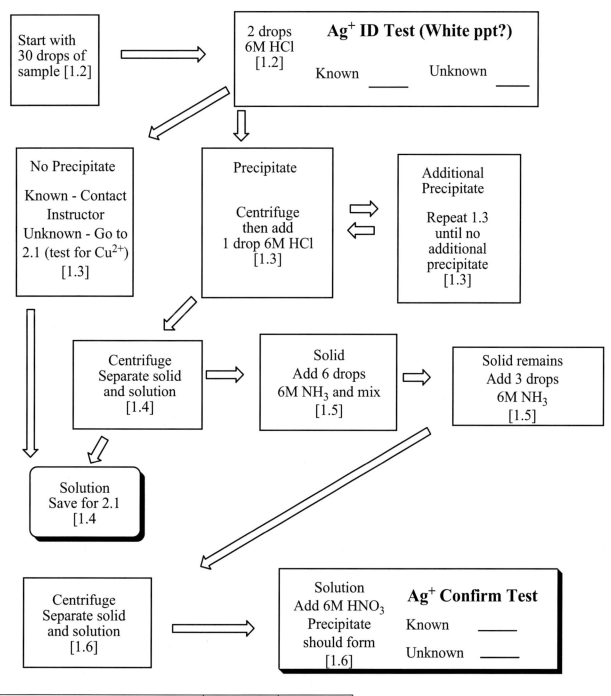

Positive Test for Silver Ion	YES	NO
Known Solution		
Unknown Solution		

Group 2: Test for Copper (II) ion (Cu^{2+})

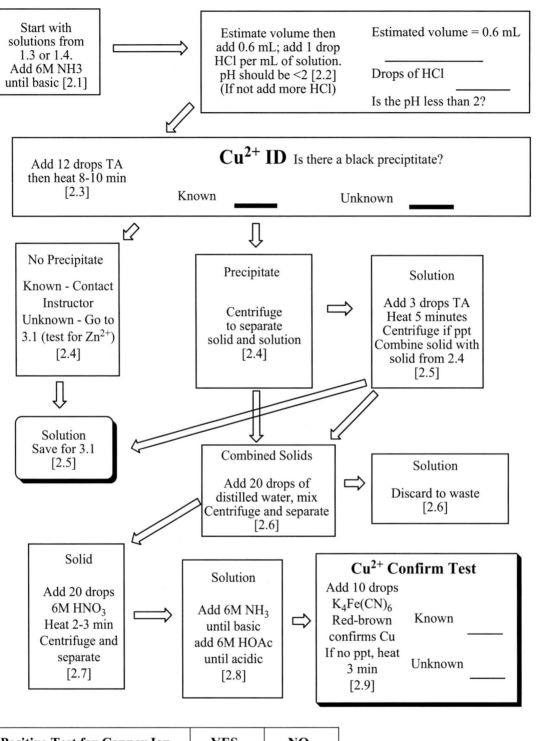

Positive Test for Copper Ion	YES	NO
Known Solution		
Unknown Solution		

Group 3: Test for Zinc (II) ion (Zn^{2+})

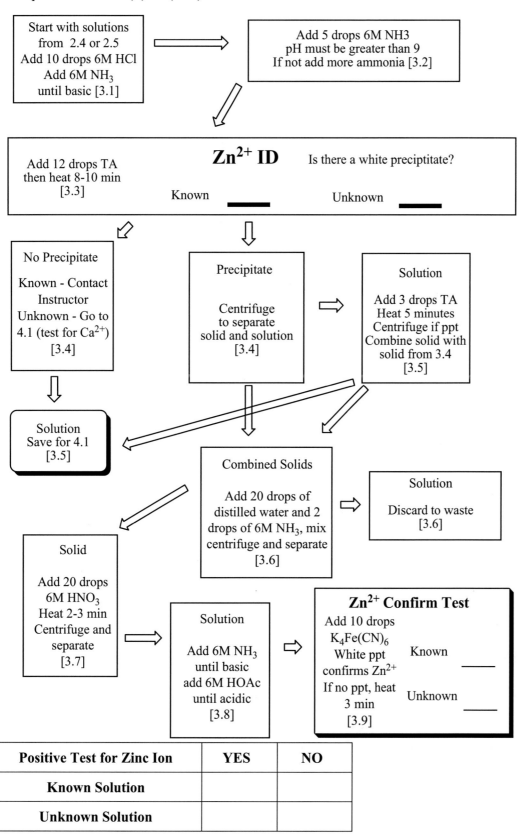

Positive Test for Zinc Ion	YES	NO
Known Solution		
Unknown Solution		

Group 4: Test for Calcium ion (Ca^{2+})

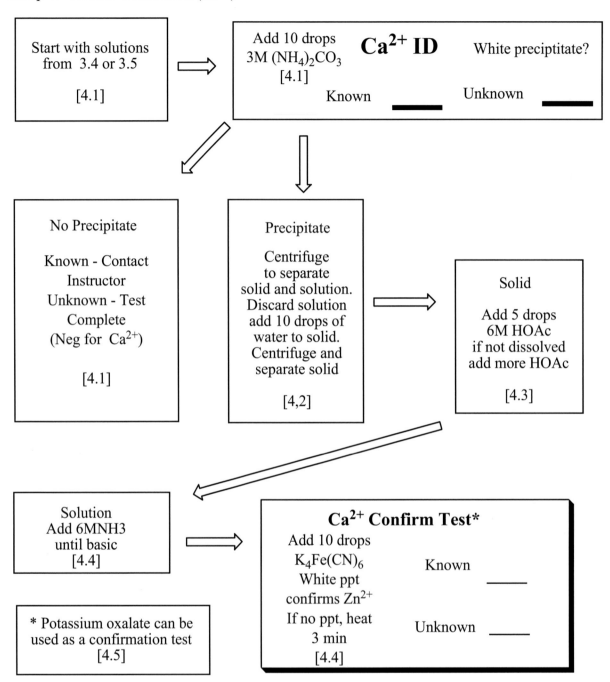

Positive Test for Calcium Ion	YES	NO
Known Solution		
Unknown Solution		

Questions:
1. Write the Ksp expression for each of the precipitation reactions used to identify each group and calculate the concentration of the anion required to precipitate a 0.005 M cation solution for each of these.

2. Using the sulfide ion concentrations from question #1 prove that it is possible to remove copper (II) ion from zinc (II) ion by adjusting the pH to 0.5 ([H^+]=0.3M). You can assume a saturated solution of H_2S (0.1M) and a combined ionization constant for H_2S dissociating into $2H^+$ and S^{2-} of 1 x 10^{-26}

3. Predict if the following conditions would result in the formation of a precipitate:
   a) 0.0025 M Fe^{3+} at pH 6.00

   b) 0.0025 M Ba^{2+} and a 1.25 x 10^{-4} M solution of sodium sulfate

   c) 0.05 M Ca^{2+} and a 1.25 x 10^{-4} M solution of cesium phosphate

   d) 7.92 x 10^{-15} M solution of Hg^{2+} and a saturated solution of H_2S at a pH of 0.5.

4. Determine the equilibrium concentration of the following:
   a) The cobalt ion concentration of a saturated solution of CoS

   b) The pH of a saturated $Zn(OH)_2$ solution

   c) The lead (II) ion concentration of a saturated solution of $PbCl_2$.

SCI 124 Principles of Chemistry II
Laboratory 10: Redox Titration
By Ching Yim

Purpose:

This laboratory introduces students to the technique of redox titration. Two techniques for obtaining the endpoint are examined: the visual technique (with and without indicator) and the potentiometric technique. The purity of a secondary reference (Mohr's salt) will be determined by titrating against a primary reference (potassium dichromate). Finally, the concentration of a solution of potassium permanganate will be determined by titration against the secondary reference.

Introduction:

Titration is one of the most important techniques to determine the concentration of a chemical compound in a solution. Beside the more common acid-base titration (which is used to determine the concentration of an acid or a base in solution), redox reactions can also be used as basis for titration – in this case, the concentration of a known oxidizing agent can be determined by titrating against the standard solution of a reducing agent, or conversely, the concentration of a known reducing agent can be determined by titrating against the standard solution of an oxidizing agent.

For any titration reaction, it is crucial to determine the endpoint of the reaction. For a redox titration, the endpoint can be determined by three different methods. First, many elements (especially some transition element) form chemical species in different oxidation states with different colors. If the difference in color is pronounced enough, the endpoint can be determined by the change in the color of the solution. For example, the transition element manganese (Mn) forms permanganate ion, a strong oxidizing agent with an intensely purple/violet color. Reduction of the permanganate ion to the pale pink manganese (II) ion (from the oxidation state of +7 to +2) produces a sharp color change that enables the endpoint to be determined visually:

$$MnO_4^-(purple) + 8H^+ + 5e^- \leftrightarrow Mn^{2+}(pink) + 4H_2O \qquad E^o = +1.51V$$

Second, the endpoint of a titration can be determined by a redox indicator (similar to the use of a pH indicator in an acid-base titration). For example, the dichromate ion, another strong oxidizing agent, can be used as the oxidizing agent in a redox titration according to the following formula:

$$Cr_2O_7^{2-}(orange) + 14H^+ + 6e^- \leftrightarrow 2Cr^{3+}(green) + 7H_2O \qquad E^o = +1.33V$$

Unfortunately, the endpoint color change is not sharp enough in practice for the endpoint to be determined precisely, so an indicator is needed. A redox indicator has a different color in its reduced form versus its oxidized form, so it can be used to indicate the completion of a redox reaction. For example, the redox indicator sodium diphenylamine sulfonate (NaDS, $E^o = +0.84V$) is purple-red in the oxidized form and colorless in its reduced form. If dichromate is titrated against the iron (II) ion:

$$Fe^{3+} + ne^- \leftrightarrow Fe^{2+} \qquad E^o = +0.77V$$

The E^o of the indicator NaDS is in between that of the dichromate ion and the product of the reaction, the Fe^{3+} ion; therefore the appearance of the purple-red color of the oxidized form of NaDS indicates the completion of the reaction and the endpoint.

Third, the endpoint of a redox titration can be determined by a potentiometer. A potentiometer uses a reference electrode (such as a standard calomel or Hg/Hg_2Cl_2 electrode) to measure the reduction potential of the solution. When the reduction potential of the solution is plotted against the volume of the titrant added, a sharp inflection point appears (akin to the inflection point that also appears in the pH versus volume plot in an acid-base titration) that marks the endpoint of the reaction. In this experiment you will be asked to use all three methods of endpoint detection mentioned above.

In this experiment you will determine the concentration of a solution of potassium permanganate. In titration, the solution of unknown concentration is titrated against a standard solution. Ideally the standard solution is made of a solid substance of known purity that is chemically stable and have no (or constant) water content in solid state, a substance called a *primary reference*. If that is not possible, the standard solution can be that of a *secondary reference*, the concentration of which can be determined by titrating against a primary reference. In this experiment, a Fe^{2+} salt {Mohr's salt, $Fe(NH_4)_2(SO_4)_2.6H_2O$} will be used as the secondary reference to determine the concentration of a $KMnO_4$ solution. The concentration of the secondary reference is in turn determined by titrating against a primary reference, potassium dichromate ($K_2Cr_2O_7$), a compound used as a standard reference by the federal government's National Institute of Standards and Technology (NIST). Note that Mohr's salt itself cannot be used as a primary reference because its water of crystallization can be lost upon prolonged storage and the Fe^{2+} ion in the salt can be oxidized by oxygen in air to become Fe^{3+}. In case you wonder, $KMnO_4$ itself cannot be used as a primary reference either, since it has a tendency to react with trace amount of organics in the environment (such as the detergent used to wash the glassware) to become manganese dioxide (MnO_2), and the concentration of $KMnO_4$ in solution tend to decrease steadily over time.

Reading Assignment

You are required to complete the following reading assignments as part of this laboratory session: General Chemistry Atoms First, *2nd Ed.*, McMurry/Fay, Chapter 7 Reactions in Aqueous Solution Section 7.10 Redox Titrations and Chapter 17 Electrochemistry.

Procedure
1.   **Observation of the color change of the redox reactions involved**

1.1   Pipet 2.0mL of 0.05M $Fe(NH_4)_2(SO_4)_2$ solution each into three separate test tubes.

1.2   Add one drop of concentrated H_3PO_4 into each tube.

1.3   Into the first tube, add 0.05M $KMnO_4$ dropwise (note: use glass droppers only). Record color change.

1.4   Into the second tube, add 0.05M $K_2Cr_2O_7$ dropwise. Record color change.

1.5   Into the third tube, add one drop of the NaDS indicator solution, then add 0.05M $K_2Cr_2O_7$ dropwise. Record color change.

2.   **Titration of Mohr's salt by potassium dichromate**

2.1   Weigh about 0.70 – 0.75g of the $K_2Cr_2O_7$ solid and record the weight to 3 decimal points.

2.2   Dissolve the $K_2Cr_2O_7$ in about 150mL distilled water.

2.3   Dilute the $K_2Cr_2O_7$ solution into 250mL by using a 250mL volumetric flask. Calculate the molarity of the $K_2Cr_2O_7$ solution (FW = 294.19 g/mol). This is your primary standard.

2.4   Weight about 2.0 – 2.5 g of the Mohr's salt $\{Fe(NH_4)_2(SO_4)_2.6H_2O\}$ and record the weight to 3 decimal points.

2.5   Dissolve the Mohr's salt in 200mL of 1M H_2SO_4. This is your secondary standard.

2.6   Pipet 25mL of the $Fe(NH_4)_2(SO_4)_2$ solution into a 100mL beaker.

2.7   Add 1.5mL of 85% H_3PO_4 into the solution and swirl gently to mix. After that, add 5 drops of the NaDS indicator into the solution. Add a magnetic stir bar and put the beaker on a stirring plate.

2.8   Use a stand and a clamp to put the calomel electrode and the reference electrode into the beaker. Connect the electrodes to the potentiometer.

2.9   Rinse your 50mL buret three times with 3mL (each) of your $K_2Cr_2O_7$ solution – remember to flush out the bubble in the tip – and then fill the buret with the $K_2Cr_2O_7$ solution. Record the volume.

2.10   Turn on your laptop and select Logger Pro.

2.11   Turn on the stirrer. Use your buret to add a 1mL increment of the $K_2Cr_2O_7$ solution to the beaker. Record the volume of the buret, the color of the solution in the beaker and the reduction potential reading on the computer screen.

2.12   Repeat the above step (2.11). If you see the change of the potential is becoming greater, add a smaller increment {0.5mL, and then 0.2mL}. The endpoint is reached when the reduction potential records a sudden large change of value. After the end point is reached, add another two 0.5mL increments of the $K_2Cr_2O_7$ solution and again, record the volume, color and reduction potential.

2.13   Now that you understand the endpoint color and the approximate volume to be used, repeat the titration twice using 25mL of the $Fe(NH_4)_2(SO_4)_2$ solution in   250mL Erlenmeyer flasks. Remember to add 1.5mL of 85% H_3PO_4 and 5 drops of the indicator. You don't need to use the electrodes and the potentiometer for these two titrations – just use the change-of-color endpoint.

2.14   Plot the titration curve (reduction potential versus total volume) for the first titration.

2.15   Calculate the concentration of the $Fe(NH_4)_2(SO_4)_2$ solution using the titration data from all three titrations and compare that to the expected concentration based on the amount of the Mohr's salt weighed out.

3.   **Titration of potassium permanganate unknown solution**

3.1   Rinse your 50mL buret three times with 3mL (each) of the $KMnO_4$ solution provided – remember to flush out the bubble in the tip – and then fill the buret with the $KMnO_4$ solution. Record the volume.

3.2   Pipet 25mL of the $Fe(NH_4)_2(SO_4)_2$ solution into a 250mL Erlenmeyer flask. Add 1.5mL of 85% H_3PO_4 into the solution and swirl gently to mix. Do **not** add the indicator.

3.3   Titrate the $Fe(NH_4)_2(SO_4)_2$ solution using the $KMnO_4$ solution in the buret. The endpoint is reach when the first drop of $KMnO_4$ turns the solution pink and do not fade away upon swirling.

3.4   Repeat the titration twice.

3.5   Calculate the concentration of the $KMnO_4$ solution using the titration data from all three titrations.

Calculations:

1.   Calculate the molarities of the $K_2Cr_2O_7$ and the $Fe(NH_4)_2(SO_4)_2$ solutions you made up.
2.   Balance the relevant redox equations {$Cr_2O_7^{2-}$ + Fe^{2+} as well as MnO_4^- + Fe^{2+}} by the half-reaction method.

3.   In the case of the potentiometric titration, plot the total volume of titrant added (V_t) on the X-axis against the reduction potential measured (E) on the Y-axis. The endpoint is the inflection point on the graph. Compare the potentiometric endpoint   with the indicator endpoint.

4.   Calculate the molarity of Fe^{2+} in section 2 by the equation
$$\frac{M_1V_1}{n_1} = \frac{M_2V_2}{n_2}$$
where M_1 and M_2 are the molarities of the reactants (in this case $Cr_2O_7^{2-}$ and Fe^{2+})
V_1 and V_2 are the volumes (in mL) of the two reactants used in the titration
a_1 and a_2 are the stoichiometric coefficients of the reactants

5.   Calculate the purity of the $Fe(NH_4)_2(SO_4)_2 \cdot 6H_2O$ solid using the molarity of Fe^{2+} calculated in the titration versus the molarity of Fe^{2+} calculated from weighing.

6.   Calculate the molarity of the $KMnO_4$ solution from your titration in section 3 similarly.

Results:

1. Observation of the color change of the redox reactions involved

Reaction	Observation
$MnO_4^- + Fe^{2+}$	
$Cr_2O_7^{2-} + Fe^{2+}$	
$Cr_2O_7^{2-} + Fe^{2+} + $ NaDS indicator	

2. Titration of Mohr's salt by potassium dichromate

Solution preparations:

Mass of $K_2Cr_2O_7$	Mass of $Fe(NH_4)_2(SO_4)_2.6H_2O$ {Mohr's salt}
Moles of $K_2Cr_2O_7$ {FW = 294.19 g/mol}	Moles of $Fe(NH_4)_2(SO_4)_2.6H_2O$ {FW = 392.13 g/mol}
Molarity of $K_2Cr_2O_7$	Expected molarity of $Fe(NH_4)_2(SO_4)_2$

Titration curve (1st experiment):

Volume added	Total volume (V_t)	Reduction potential (E)	Color

Graph (E versus V_t):

Titration data:

Titration	1st trial (color)	1st trial (curve)	2nd trial	3rd trial
Initial volume (mL)				
Endpoint volume (mL)				
	Average:		N/A	N/A
Volume of titrant used (mL)				

Average volume:

Balance the redox equation of $Cr_2O_7^{2-} + Fe^{2+} \rightarrow Cr^{3+} + Fe^{3+}$

Calculate the molarity of Fe^{2+} (represented by M_2):

$$\frac{M_1V_1}{n_1} = \frac{M_2V_2}{n_2}$$

Remember M_1 and M_2 are the molarities of the reactants (in this case $Cr_2O_7^{2-}$ and Fe^{2+})

V_1 and V_2 are the volumes (in mL) of the two reactants used in the titration

n_1 and n_2 are the stoichiometric coefficients of the reactants

Percent error of the concentration of $Fe(NH_4)_2(SO_4)_2$:

$$= \frac{[Fe^{2+}]_{titration} - [Fe^{2+}]_{weighing}}{[Fe^{2+}]_{weighing}} \times 100\%$$

$=$

3. Titration of potassium permanganate unknown solution

Titration	1st trial	2nd trial	3rd trial
Initial volume (mL)			
Endpoint volume (mL)			
Volume of titrant used (mL)			

Average volume:

Balance the redox equation of $MnO_4^- + Fe^{2+} \rightarrow Mn^{2+} + Fe^{3+}$

Calculate the molarity of MnO_4^-:

$$\frac{M_1V_1}{n_1} = \frac{M_2V_2}{n_2}$$

<u>Questions:</u>

(1) From your observation of the $Cr_2O_7^{2-}$ versus Fe^{2+} reaction (section 1), is it possible to titrate $Cr_2O_7^{2-}$ against Fe^{2+}? Why?

(2) Why do you think that Fe^{2+} is put into the flask to be titrated by MnO_4^- in the buret and not the other way around? {Hint: think color detection.}

(3) Both the titrations require the presence of acids in the solutions (in both cases, we use 1M H_2SO_4). Explain. Could you use HCl instead? Why?

(4) In all of the titrations, H_3PO_4 is used as an additive. Explain the helpfulness of using H_3PO_4 by noting the reaction $Fe^{3+} + HPO_4^{2-} \rightarrow Fe(HPO_4)^+$ and by using the Le Châtelier's Principle you learnt in Chapter 14. {Hint: is Fe^{3+} one of the reactants or one of the products? And where does the HPO_4^{2-} come from?}

SCI_124 Principles of Chemistry II

Laboratory 11: Electrochemistry

(based in part on Small Scale Chemistry methodology as described in Chemtrek by Stephen Thompson at Colorado State University)

Purpose:

This laboratory examines electrochemical cells and electrochemical processes. The difference between a voltaic and electrolytic cell is presented and electrolysis is performed. A series of voltaic cells are measured and compared to reference values.

Introduction:

The topic of oxidation-reduction reactions has already been covered in the SCI123/SCI124 course sequence. In those discussions oxidation-reduction reactions were characterized as reactions in which there is a <u>transfer of electrons</u>. The number of electrons around an atom in a structure is reflected in the oxidation number of that atom. The working definition of an oxidation-reduction reaction that we have been using is that an oxidation-reduction reaction is any reaction in which there is a change in oxidation numbers. When oxidation numbers increase the material is said to be oxidized (one example would be

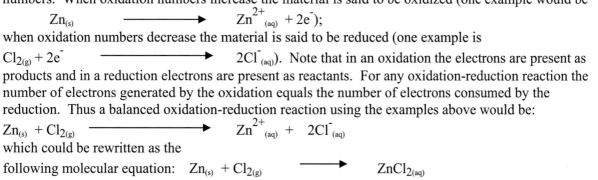

$$Zn_{(s)} \longrightarrow Zn^{2+}_{(aq)} + 2e^-);$$

when oxidation numbers decrease the material is said to be reduced (one example is

$$Cl_{2(g)} + 2e^- \longrightarrow 2Cl^-_{(aq)}).$$ Note that in an oxidation the electrons are present as products and in a reduction electrons are present as reactants. For any oxidation-reduction reaction the number of electrons generated by the oxidation equals the number of electrons consumed by the reduction. Thus a balanced oxidation-reduction reaction using the examples above would be:

$$Zn_{(s)} + Cl_{2(g)} \longrightarrow Zn^{2+}_{(aq)} + 2Cl^-_{(aq)}$$

which could be rewritten as the

following molecular equation: $Zn_{(s)} + Cl_{2(g)} \longrightarrow ZnCl_{2(aq)}$

Electrochemistry is the observation of oxidation-reduction reactions using some sort of electrical circuit. Since electrons are being generated by the oxidation and consumed by the reduction, if we can physically separate the oxidation from the reduction and connect them together using wires (which will pass only electrons) we can measure common electrical quantities such as current and voltage. In fact when you use any type of battery the voltage (and when in use the current) is supplied by oxidation-reduction reactions occurring inside the battery. In this laboratory we will measure current and voltage for some chemical reactions. The student will also gain an appreciation of what reactions are spontaneous and what are not spontaneous. The generation of standard reduction potentials and how to use these values will be performed by the student.

You are familiar with oxidation-reduction reactions that occur when the reactants are mixed which means that they occur spontaneously. One example would be the oxidation of iodide ion to iodine by persulfate which you used in the kinetics lab. If you can physically separate the oxidation from the reduction reaction and connect wires (also called an electrical lead) to each so that the electrons will flow you will observe that the common characteristic of all of these reactions is that they would exhibit a positive voltage (this is called an electrochemical cell). **<u>An electrochemical cell which exhibits a positive voltage is called a voltaic cell and represents a spontaneous chemical reaction.</u>**

Conversely if you connect the electrical leads and obtain a negative voltage, the reaction will not occur spontaneously. These types of reaction will occur only when electrical energy is supplied. **An electrochemical cell which exhibits a negative voltage is called an electrolytic cell and represents a non-spontaneous chemical reaction that will occur only when electrons are supplied from an external source.** If you have two cells present and connect the wires to observe a voltage it will be positive for the spontaneous reaction and if you reverse the positive and negative leads the voltage will display the same negative value. That means that any electrochemical cell will be voltaic (spontaneous, positive voltage) with the leads set up in one direction (+/-) and electrolytic (non-spontaneous, negative voltage) when the leads are reversed.

You will be performing three types of experiments in this laboratory session:

1. The first is an electrolysis which means supplying electrical current to a non-spontaneous electrochemical cell (electrolytic cell) to cause the oxidation-reduction reactions to occur. You will take two strips of copper metal and place them in a solution of copper ion. In the absence of an electrical current there will be no chemical change. If a voltage is placed across the copper strips oxidation will occur at one copper strip and reduction at the other. When you supply a voltage to the electrodes (copper strips) you force a non-spontaneous process to occur. You will measure the chemical changes observed and calculate the current the flowed through the cell.

2. The second is an examination of what is considered to be a spontaneous and non-spontaneous process. You should observe that some reactions occur when you combine the materials together. These are spontaneous processes and if you measured the voltage of such an electrochemical cell the voltage would be positive. Some oxidation-reduction reactions will occur only when a voltage is applied (in this case by using a 9V battery). The corresponding electrochemical cell in these cases would be electrolytic.

3. You will measure the voltage of a number of electrochemical cells and record the values. You will then prepare a table for the reduction of each material relative to a standard reduction (copper ion to copper metal). You will then adjust these values to match the standard reduction potentials in your text book (Chang, Table 19.1, this requires that the reduction of hydrogen ion to hydrogen gas be equal to zero). The values that you obtain will be compared to the values found in the table.

Reading Assignments:

You are required to complete the following reading assignment as part of this laboratory session: General Chemistry: Atoms First, *2nd Ed.*, McMurry/Fay, Chapter 17 Electrochemistry.

Grading:
Grading will be as described in the general laboratory handout.

Procedure:

General: You should keep a 600 mL beaker at your workstation to use as a satellite waste receptacle.

1. Electrolysis

1.1 Obtain the mass of two copper strips on an analytical balance to within 0.1 mg. Mark each of the strips or keep them separate such that you know which strip will be used for the negative and positive post in the electrolysis (cathode and anode respectively).

1.2 Place the two copper strips at the opposite sides of a 50 mL beaker or use a larger beaker if this size is not available. Turn the ends of the strips over the top of the beaker and add 30 mL of 1M $CuSO_4$ solution.

1.3 Connect the electrical leads to the copper strips **WITHOUT CONNECTING THE ELECTRICAL LEADS TO THE VOLTAGE SOURCE.**

1.4 Once you are certain that the connections are secure connect one post at a time to the voltage source. If any problems arise after the initial connection or during the electrolysis disconnect one of the electrical connections from the voltage source.

1.5 Make a note of which copper strip was connected to the negative post of the voltage supply and which was connected to the positive post.

1.6 Allow the electrolysis to proceed for 15 minutes and once the 15 minutes has elapsed remove the connections from the voltage source first.

1.7 Remove the two copper strips and return the copper sulfate solution to the container provided. Rinse the copper strips with distilled water and dry them with a paper towel.

1.8 Determine the mass of each of the copper strips on an analytical balance and record this data on the data sheet provided. Be sure that you know which copper strip was connected to which post on the voltage source.

2. Qualitative evaluation of spontaneous and non-spontaneous oxidation-reduction reactions

2.1 Using the a 24 well plate add approximately half fill the first well with 1M $CuSO_4$ solution half fill the second well with 1M $Zn(NO_3)_2$.

2.2 Place a small piece of zinc metal in the copper sulfate solution and a small piece of copper into the zinc solution. Let the metals sit in the solution for at least 1 minute. Remove the two pieces of metal and record any observations.

2.3   Take the zinc metal which was dipped in the copper solution and connect it to the positive post of a 9V battery using the alligator clips provided. Connect a graphite pencil lead to the negative post in a similar way and place both solids in the copper sulfate solution without allowing the solids to touch one another. Allow the solids to remain in the solution connected to the battery for 2 minutes. Remove the electrodes and rinse with distilled water and record your observations.

2.4   Repeat 2.3 using the zinc nitrate solution with the copper attached to the negative post of the battery and the graphite pencil lead attached to the positive post. Record your observations after the 2-minute period.

3. Measurement of voltaic cell potentials.

3.1   Connect the iBook to the LabPro box using a USB Cable. Connect a power supply to the LabPro box and connect the voltage leads into Channel 1. Turn on the computer and open Logger Pro (the caliper symbol on the desktop) program. The computer should display a value for voltage. Connect the two voltage probes (red and black) together and confirm that the voltage is near zero. Separate the leads and observe the value.

3.2   Using strips of filter paper place drops of the following 1M solution around the edges of the strips of filter paper and mark which solution is which using a pencil: $CuSO_4$, $Zn(NO_3)_2$, $AgNO_3$, $PbSO_4$, $FeSO_4$.

3.3   Place one of each of the small solid pieces on the corresponding solution and connect each of the solution drops to each other through the center using drops of the 1M KNO_3 solution.

3.4   Place the electrode leads on copper and one of the other metals. Record the reading if it is positive and mark on the data sheet which metal is connected to the positive lead and which is connected to the negative lead. If the reading is negative reverse the leads and record as above.

3.5   Repeat 3.3 for all of the metals on the filter paper strips recording the voltage and which metal is positive and which is negative.

4. Calculations and comparison with standard values

4.1   For the electrolysis experiment calculate the weight gain and weight loss of each of the copper strips. Convert each mass into moles and each number of moles of copper into moles of electrons.

4.2   One mole of electrons carries of charge of 9.65×10^4 Coulombs (one Faraday). Calculate the amount of charge transferred in the 15 minute period by converting the number of moles of electrons into Coulombs (C).

4.3   Since an ampere (A), which is a unit of current, is a Coulomb/second (C/s) calculate the total current passed through the cell for both electrodes.

4.4   Determine the % difference between the current obtained from one electrode to that for the other electrode.

$$\frac{\text{Value 1 - Value 2}}{\text{Avg of Value 1 + 2}} \times 100\% = \% \text{ difference}$$

4.5   In order to compare the values you obtained from the voltaic cells to standard values you will need to report all of the values as **reduction potentials**. This means the values will be positive only if the non-copper metal was connected to the positive post. Use Zn and Ag on the table to be sure you have the sign convention correct; zinc should be a negative standard reduction potential and silver should be a positive standard reduction potential.

4.6   Order you data as reduction potentials in decreasing order (remember Cu/Cu(II) will be zero since that was the reference used).

4.7   Since standard tables use hydrogen as a reference material rather than copper, add 0.34V to each of the values to convert from a copper reference to a hydrogen reference. Find the corresponding reference values in the McMurray/Fay textbook Chapter 17 Table 17.1 or in Appendix D and record those values in the table provided.

4.8   Calculate the % error between the values you obtained and those in the textbook. Enter these values into the table provided.

$$\% \text{ error} = \frac{\text{Expt. value - Reference value}}{\text{Reference value}} \times 100\%$$

Results:

1. Electrolysis of Copper

	Positive Post (Anode)	Negative Post (Cathode)
Mass of Copper Electrode Before Electrolysis		
Time of Electrolysis (minutes)		
Mass of Copper Electrode After Electrolysis		
Mass change of Copper Electrode		
moles of Copper consumed/generated		
Charge transferred during Electrolysis		
Total Current passed through each electrode		
% Difference in current values		

2. Qualitative evaluation of spontaneous and non-spontaneous reactions

Zn metal in $CuSO_4$ soln	
Zn metal/Graphite/ 9V Battery ($CuSO_4$ soln)	
Cu metal in $Zn(NO_3)_2$ soln	
Cu metal/Graphite/ 9V Battery ($Zn(NO_3)_2$ soln)	

3. Voltaic cell potentials (raw data, 3.3-3.4)

Voltage (must be positive)	Positive Terminal (+)	Negative Terminal (-)

Reduction Potentials (4.5-4.8)

Metal	Reduction Potential (4.5) (Cu reference)	Reduction Potential (4.7) (Hydrogen reference)	Reduction Potential from Table (4.7; Chang Table 19.1)	Percent error between values (4.8)

Questions:

1. Provide a brief definition of the following terms:

a) Cathode

b) Anode

c) Standard Electrode Potential

d) Voltaic Cell

e) Electrolytic Cell

2. Provide the chemical reaction of an oxidation-reduction reaction that was observed to be spontaneous. Provide a value for the standard cell voltage for this reaction. Is the voltage a positive value as expected for a spontaneous chemical reaction?

3. If the reaction in part 2 with the 9V battery in the zinc nitrate solution produced zinc metal at the cathode and oxygen gas at the anode, provide the chemical reaction and a value for the standard cell potential expected for this electrochemical cell. Was this reaction spontaneous and does the standard cell potential calculated agree in sign (positive or negative) with what you observed for this reaction? Explain briefly.

4. For the voltaic cells that you measured that used silver and copper provide the reaction that represents the voltage observed. Repeat this for the cell that used lead and copper. Does copper behave in these two cells in the same way? Is this what you would expect from the standard reduction potentials for the reduction of silver (I) ion, copper (II) ion and lead (II) ion?

5. Using standard cell reduction potentials indicate the voltage expected for each of the following cells under standard conditions. Label each cell as voltaic or electrolytic. Also indicate which, if any, of the cells below would need to have a electrode material represented in the cell diagram to give a working electrochemical cell.

a) $Cr_{(s)} | Cr^{3+}_{(aq)} || Hg_2^{2+}_{(aq)} | Hg_{(l)}$

b) $I^-_{(aq)} | I_{2(s)} || Cr_2O_7^{2-}_{(aq)} | Cr^{3-}_{(aq)}$

c) $Al_{(s)} | Al^{3+}_{(aq)} || F_{2(g)} | F^-_{(aq)}$

d) $Br^-_{(aq)} | Br_{2(l)} || Cl_{2(g)} | Cl^-_{(aq)}$

e) $Sn_{(s)} | Sn^{2+}_{(aq)} || Mg^{2+}_{(aq)} | Mg_{(s)}$